HAZARD IDENTIFICATION OF

Ad van Dommelen

HAZARD IDENTIFICATION OF AGRICULTURAL BIOTECHNOLOGY

Finding Relevant Questions

International Books, 1999

This work was financially supported by the Netherlands Organization for Scientific Research (NWO), under grant # 225-95.007

ISBN 90-5727-034-X
Keywords: agriculture, biosafety, biotechnology, controversy, familiarity, genetic engineering, hazard identification, technology assessment, precautionary principle, risk

Cover design: Marjo Starink
Desk Top Publishing: Hanneke Kossen
Printing: Drukkerij Haasbeek

International Books, A. Numankade 17, 3572 KP Utrecht, the Netherlands,
tel. +31 30 2731840, fax +31 30 2733614, e-mail: i-books@antenna.nl

When the Master went inside the Grand Temple, he asked questions about everything. Someone remarked, 'Who said that the son of the man from Tsou understood the rites? When he went inside the Grand Temple, he asked questions about everything'. The Master, on hearing of this, said, 'The asking of questions is in itself the correct rite'.
(The Analects of Confucius, Book III)

To my Mother,
In loving memory of my Father

Biographical Note

Ad van Dommelen (1961) studied philosophy and science dynamics at the *University of Amsterdam* and the *New School for Social Research* in New York. After graduating in 1988 he was a teacher and researcher in the *Department of Applied Philosophy* at the *Agricultural University of Wageningen* until 1992. From then on he received a research grant from the *Netherlands Organization for Scientific Research* (NWO) to write his PhD thesis in the *Department of Theoretical Biology* at the *Free University of Amsterdam.* From 1990 until 1993, he was an editor of *Zeno,* journal for sience, technology and society. In 1995 he took the initiative to organize four international workshops and to be the editor of the resulting, *Coping with deliberate release: The limits of risk assessment* (1996). Since 1998 he is working as a business partner in *Houtsma & van der Schot,* office for research, consultancy and journalism in the field of environment and economy. He can be reached at Keizersgracht 8, 1015 CN Amsterdam, phone: +31.20.6239480, fax: +31.20.6388771, e-mail: hout.schot@inter.nl.net.

Contents

List of Abbreviations and Acronyms

ABRAC	Agricultural Biotechnology Research Advisory Committee
ACRE	Advisory Committee on Releases to the Environment
APHIS	Animal and Plant Health Protection Service
CCRO	Coordination Commission Risk-Assessment Research
CEC	Commission of the European Communities
CFC	Chlorofluorocarbons
COGEM	Commissie voor Genetische Modificatie
COP	Conference of the Parties
DDT	Dichloro-Diphenyl-Trichloro-ethane
DNA	Deoxyribonucleic Acid
EPA	Environmental Protection Agency
ESA	Ecological Society of America
EU	European Union
FOE	Friends of the Earth
GAO	General Accounting Office
GEM	Genetically Engineered Microorganism
GEO	Genetically Engineered Organism
HR	Herbicide-Resistance
IBD	Institut für Biochemie Darmstadt
NAS	National Academy of Sciences
NGO	Non-governmental Organization
NIH	National Institutes of Health
OECD	Organization for Economic Cooperation and Development
ÖIF	Öko-Institut Freiburg
PEER	Public Employees for Environmental Responsibility
RAC	Recombinant DNA Advisory Committee
SRQ	Set of Relevant Questions
TA	Techology Assessment
TAB	Büro für Technikfolgen-Abschätzung des Deutschen Bundestages
UNEP	United Nations Environmental Programme

UCS	Union of Concerned Scientists
USDA	United States Department of Agriculture
WWF	World Wildlife Fund
STOA	Scientific and Technological Options Assessment

Preface

Where the products of science and technology become more and more sophisticated, the assessment of their impact on our lives may increasingly become a privilege of experts. Standing at the brink of entering the 21st century, we are faced with the challenge of managing a *high tech* future.

The book in your hands is an attempt to see through the wall of *expertocracy* that may obstruct our outlook on the process of technology assessment. It presents a close reading of some of the scientific controversies that seem to stand in the way of a balanced assessment of modern biotechnology.

The framework of reflection developed in the following pages aims to interpret a host of very diverse scientific and political literature in its context of connections. The interplay between science, technology and society has impressive intricacies, which pervade the 'biotechnology debate' at all its levels of concern.

Disagreement in science is undoubtedly an asset for the development of new ideas, but it may pose a problem when it is found at the basis of scientifically informed policy choices about a wise approach to attractive new possibilities. Societies can hardly make such choices without a practical perspective on the status and structure of controversies in applied science. This need for a pragmatic interpretation of scientific disagreements will become more pressing with a further sophistication of the products of science and technology as they are presented to society.

Technology assessment has always been part of the development of humankind and has probably always been subject to debate. For societies that wish to secure their *sustainable development* in the future, it should remain a daily concern. The processes of learning that have brought us the fruits of modern technology must be parallelled by the processes of learning that allow us to apply these fruits with due prudence.

Many people have been constitutive to the genesis of this book. I gratefully acknowledge all who made me think one step further in the numerous discussions, criticisms, and feedbacks that paved the road of my research. Among them were my colleagues in the *Department of Theoretical Biology* at the *Free*

University of Amsterdam. Special thanks for their special support goes to Hugo van den Berg, James Goodwin, Wim Heiko Houtsma, Harro van Lente, René von Schomberg and Jos van der Schot. The intellectual challenges that were posed to me through the years by Tjard de Cock Buning, Philip Regal and Wim van der Steen have been decisive for the development of the ideas in this book. Most of all, I thank my friends for being my friends – without them this work would not have been done.

Amsterdam, Spring 1999

CHAPTER 1

Scientific Controversy in Biosafety Assessment

— 1.1 Contested expertise on risks of biotechnology —

New technologies raise new questions about possible future effects. The challenge for technology assessors is to uncover the relevant questions for the assessment of a specific technology. As a preliminary example, consider the so-called *millennium problem* of computer software. The cause of the problem is so surprising in its simplicity that it feels like the plot of a best-selling science fiction novel rather than as a real world problem. Computer programmers have failed to adequately address a simple question: what could be the long-term consequences if year-counts are represented in a two-digit format rather than in a four-digit format? In times when computer memory was more expensive than it is now, programmers chose to be as thrifty as possible and to use two digits rather than four to represent a year-count: the year *nineteen seventy-five*, for example, would be represented as '75' instead of as '1975'. Back then they did not stop to ask about the possible consequences if some of these computer programmes would still be used in the eve of the year 2000.

Now, even in a small, albeit technologically advanced, country as the Netherlands, it is estimated that the costs to 'repair' this problem will rise to about 20 billion Dutch guilders (approximately 10 billion US dollars). And, possibly, the costs will rise to a higher figure after the year 2000 has started. If early computer programmers had thought of asking the question 'what will happen when we pass the year 99?', they would most probably have decided to accept the extra cost of using more computer memory for two more digits to express the year-count. Failure to adequately address a simple question in the early assessment of a now widespread technology is confronting societies with high costs today and may lead to still higher, at the time of this writing as yet unforeseeable, costs.

Modern biotechnology raises new questions about its possible future effects, as other new technologies do. Genetic engineering brings along new questions about a range of concerns such as: *public participation*, *sustainability*, *insurance*, *patenting, labelling, environmental law, ethics* and *biosafety*. This study specifically deals with questions in relation to the *biosafety of genetic engineering*. In this introductory chapter, the general research problem of this study and the choices that have led to the presented approach of this research problem are specified.

1.1.1 Science for policy

Biotechnology companies are eager to bring their products to the market. Politicians face the challenge of promoting the possible benefits and at the same time limiting the possible costs that genetic engineering may have in stock for society. Policy-makers around the world are expected to address the relevant questions in relation to possible unwanted costs of this new technology and to use these questions as a basis for prudent regulation of its applications. The scientific community is a vital source of expertise concerning those relevant questions on the biosafety of genetic engineering. As will be shown, the scientific community has difficulty reaching consensus about an adequate way of approaching the biosafety assessment of genetic engineering (cf. van Dommelen 1996c).

If scientific experts are in disagreement about a highly technical subject such as the possible risks of modern biotechnology, then where does that leave a truth-seeking policy-maker? An official working for the European Parliament's *Committee on the Environment* and a member of the STOA Programme Team (Scientific and Technological Options Assessment for the European Parliament), found himself in this position when he was asked to think about the prerequisites of adequate regulation for the contained use and deliberate release of genetically engineered organisms (Lake 1991). In the research process he came across conflicting scientific views on the matter and found reason to distinguish between what he called "orthodox experts" and "unorthodox experts", and asked himself the question: "... how are we to make sense of these contradictory forces, and the different cultural assumptions which underlie the various 'expert opinions' which politicians are faced with?" (Lake 1991: 14).

This policy-maker decided that maybe a practical strategy to address possible risks of applying genetically engineered organisms, would be to consult the *free market of insurances*. "Accordingly", Lake reports, "Lloyds of London were informally contacted, as the acknowledged centre of the insurance world's 'unusual risks' market, and asked if the deliberate release of genetically modified organisms was an insurable risk. After a few days thinking about it, they informally communicated the view that, although they did not know what the premiums would be, so far as they could see this was an insurable risk" (Lake 1991: 12).

How acceptable would this approach be as a basis for regulatory policy? Politicians are not very well prepared to deal with policy-relevant controversies in science. Lake points out that in the European Parliament only a minority of 4% of the members has a background in science (Lake 1991: 10; Lake 1989), which makes it even more difficult for the representatives to judge the implications of scientific controversies for political decisions. The situation described by Lake in relation to the biosafety of genetic engineering has parallels with the present political and scientific debate on the possible existence of an enhanced 'greenhouse' effect as a result of CO_2 exhausts, raising concern about a possible 'global warming' (cf. Michaels and Knappenberger 1996).

Both technologies (genetic engineering and uses of fossil fuels leading to CO_2 exhausts) are human activities with possible impacts on society and nature about which scientific experts are not in agreement. This leaves politicians with the problem of developing policy against a contested scientific background. Worldwide there are many politicians and policy makers concerned with issues of biosafety and other science-based decisions, who find themselves in the same awkward position as described by Lake. This policy challenge of deciding against a background of contested science should be a concern for scientists as much as it is a concern for policy-makers and for democracies in general.

ONGOING CONTROVERSY OVER BIOSAFETY ASSESSMENT...

"New regulations in dispute" (Wright 1986)

"Transgenic Plants on Trial" (Kareiva 1993)

"Trouble in the wind over altered soya beans" (MacKenzie 1995)

"Europe halts march of supermaize" (Coghlan 1996)

"Modified maize faces widening opposition" (MacKenzie 1997)

"Who's afraid of genetic engineering?" (Carter 1998)

"Seeds of disaster" (HRH Prince of Wales 1998)

FIGURE 1.1: Selection of quotes illustrating how controversies have pervaded debates on the biosafety of genetic engineering down to its very pores.

The most encompassing level at which ongoing controversies over the biosafety assessment of genetic engineering now have impact on international policy is in the negotiations about a 'biosafety protocol' in meetings of the *United Nations Environmental Programme* (cf. Dickson 1995; Regal 1995a, Regal 1995b; Independent Group 1996; Auken 1996). Recent publications such as *Grüne Gentechnik im Widerstreit* (van den Daele *et al.* 1996) and *Policy Controversy in Biotechnology* (Miller 1997) indicate that the problem of contested biosafety assessment has not been resolved (see Figure 1.1).

1.1.2 Aim and scope of this study

When countries of the European Union reach no consensus on the expected biosafety of releasing a genetically engineered organism, Article 21 of Directive 90/220 *On the deliberate release into the environment of genetically modified organisms*, requires that the European Council, "shall act by a qualified majority" (CEC 1990: 21). This means that scientific controversies may eventually be subjected to a political vote, as was done in the case of an application for the release of herbicide-tolerant oilseed rape (cf. Levidow *et al.* 1996; see also von Schomberg 1998a, 1998b). When politicians must base their decisions on contested scientific expertise, a society should at least strive to spell out the involved scientific controversies as clearly as possible. Failing to do so may lead to undemocratic situations in which political decisions are presented in an unjustified scientific guise.

If politics and science cannot be *separated* in practical decisions like developing biosafety regulation for genetic engineering, then democratic politics requires us to at least *distinguish* their respective bases of legitimation. Biosafety controversies are often riddled with terms and concepts which amalgamate scientific claims and political choices. Scientific experts take refuge to the use of arguments that are based on *analogy*, *similarity*, *plausibility* and *familiarity*. In the end, these lines of reasoning may do more to enhance antagonism over the biosafety of genetic engineering than to resolve it. Without proper clarification such arguments can become like *Trojan horses*, carrying an unexpected and unwanted load into the debate.

The general problem that is studied in this book is the occurrence of *controversy in science for policy*. The specific focus of this study is on *controversies over biosafety assessment of genetic engineering*. The present research is an attempt to develop a fundamental *and* pragmatic approach to this pressing problem. Since controversies in applied science will typically involve more than one scientific discipline and will often affect scientists as well as nonscientists, it is a requirement that the presented framework of analysis is both accessible for diverse parties in the debate and practically applicable to the issues of discussion. This study aims at developing such a framework and at demonstrating how it can make a difference to past, present and future approaches of controversies over the biosafety assessment of genetic engineering.

Summing up, the main objectives of this study are:

- To develop a *methodological analysis of controversy in science for policy* as a basis for evaluating conflicting scientific claims,
- To apply the developed analytical framework in the *evaluation of ongoing controversies over biosafety assessment of genetic engineering*,
- To have the framework be *fundamental* in the sense that it goes to the core of the involved scientific research,
- To have the framework be *accessible* to many parties in the sense that it can be understood and applied from different perspectives,
- To have the framework be *pragmatic* in the sense that it gives a constructive tool to participants in the debate,
- To demonstrate how the developed analytical framework can make a *significant difference to past, present and future approaches of controversy in science for policy*.

These objectives require a reconstruction of the science as distinct from the politics of the debate. To be successful, the developed analysis must allow an articulation of the scientific basis of the frequent references to 'analogy', 'similarity', 'plausibility', or 'familiarity', which are used as appealing 'thought-bridges' in the current debate on biosafety assessment. They serve well to forge intuitive connections between what is known and what is to be known, but in the process they may stand in the way of a genuine scientific understanding of the problem at hand.

A proper distinction of the *science* and the *politics* involved in biosafety controversies will not by itself resolve the issues, but it will help us do proper science and proper politics in the respective arenas of legitimation. Scientists find their legitimation in methodology and experiment; politicians find their legitimation in democratic procedures. Differences between the two should not be amalgamated in policy procedures and should remain separately visible in the process of decision-making on the introduction of new technologies. If it is possible to bring more clarity in discussions over the biosafety assessment of genetic engineering, then this may contribute to the more encompassing cultural debate over the prominence that societies want to give to modern biotechnology.

The possible impact of genetic engineering on societies and natural environments is a cause for concern. Especially, since those affected by the costs may not be among those who profit from the benefits (cf. Rehmann-Sutter and Vatter 1996, Rehmann-Sutter 1998). To help decision-makers deal with controversial scientific knowledge – and draw information and wisdom from "orthodox" as well as from "unorthodox" expertise – a practical tool for analysis and evaluation of scientific controversies in the context of biosafety assessment needs to be developed. The present study demonstrates how, in specific cases, controversy in applied science can be taken away by proper methodological analysis and can thus be taken out of the way of politicians involved in processes of decision-making based on policy-relevant science.

The political impact of debates about genetic engineering has been referred to as "molecular politics" (Wright 1994) and "biopolitics" (Shiva and Moser 1995). Since there is no political way *around* scientific controversies, they must be faced head on and an attempt should be made to understand them from the *inside* of science to safeguard the scientific basis of decision-making. The practical challenge of this study is: *what would be a scientific basis for biosafety regulation of genetic engineering?* The fact that many nonscientists are involved in the larger decision-making processes adds to the urgency of developing a clear and accessible analysis of the structure of scientific controversy in a policy context. This requires a pragmatic *and* fundamental perspective on contested scientific claims in the context of applied science, such as exist in biosafety assessment.

In this introductory chapter, the general subject of this research is explicated by addressing two questions: *what role can science play in the process of biosafety assessment?* (Section 1.3) and *what can be done when we find applied science in controversy?* (Section 1.4). Both questions pertain to the core of policy decisions about a vital realm of human culture, touching upon social, economic, environmental and health aspects of individuals and societies: *modern agriculture* (cf. Doyle 1986). In subsequent sections of the present chapter, a preliminary introduction (Section 1.5) and a practical illustration (Section 1.6) of the general approach taken to these questions is presented.

In the following chapters, a more detailed analysis of ongoing biosafety controversies is presented, applied and evaluated. Throughout this study many contested scientific claims in the biosafety debate will be discussed and analysed. In Chapter 2, an analytical framework is presented as a pragmatic tool to interpret and understand controversies in applied science. In Chapter 3, the presented approach will be used to demonstrate its different practical consequences in comparison to past and existing approaches to the biosafety assessment of genetic engineering. In Chapter 4, four categories of recurring biosafety controversies are analysed and evaluated through the use of the framework developed in Chapter 2. In Chapter 5, the potential for practical use of this framework in present and future biosafety regulation is specified. Finally, in Chapter 6, the consequences of the presented analytical framework for the associated responsibility of scientific experts is evaluated. Together these chapters shed a new and constructive light on a number of disputes that have appeared to be irresolvable.

— 1.2 Transgenic organisms and biosafety —

In 1973, a groundbreaking scientific discovery was made that would open the road to recombinant DNA technology (or genetic engineering) and a new era of biotechnology. Scientists successfully transferred DNA from one life form into another. Stanley Cohen and Herbert Boyer were the first to demonstrate the possibility of recombining the building-stones of the DNA-molecule. They managed to recombine ('splice') sections of viral DNA and bacterial DNA, which had been cut with the same restriction enzyme, creating a plasmid with dual antibiotic resistance. They then spliced this recombinant DNA molecule into the DNA of a bacterium, thereby producing the first *transgenic* organism. This new technique opened up the potential to retrieve a gene coding for a certain characteristic from one organism and transporting it to and bringing it to expression in another organism.

Shortly after the new recombinant DNA technique became available, the scientific community took a unique initiative. In a conference that was held at Asilomar Center in Pacific Grove, California in 1975, concern was raised about the possibility that the new technology might lead to inadvertent risks (cf. Berg *et al.* 1975). The main issue then was the risk of laboratory use and contained applications of the new technique (cf. Regal 1996). One specific risk issue that was raised in this context, for example, was the question whether experiments with genetically modified *Escherichia coli* (a normally harmless gut bacteria) would pose a threat to laboratory workers, or not. The scientists that were convened called on the US *National Institutes of Health* (NIH) to oversee this new technology. This led to *Recombinant DNA Advisory Committees* (RACs) and Guidelines (cf. Miller 1997: 2).

Nobel prize winner James Watson later remarked about this period that the effort to assess and control genetic engineering was, "a massive miscalculation in which we cried wolf without having seen or heard one" (cit. in Wright 1994: 254; see also Watson 1981). On the other hand, another Nobel prize laureate from the field of molecular biology, Erwin Chargaff (who discovered the equal nucleotide ratios in DNA, a stepping-stone towards the discovery of the structure of DNA by Watson and Crick), has arrived at a very different conclusion: "In manipulating processes worked out by nature in the wisdom of millions of years one must be aware of the danger that our shortcuts may carry a bleeding edge" (Chargaff 1987: 199). It shows that learned men with as good a grasp on the matter as can be expected, may still find themselves in opposing positions.

The controversy then born is still alive, albeit in a different guise. A recent illustration of the ongoing dispute over biosafety of genetic engineering is the 'Statement' that was issued by a group of scientists, "concerned about current trends in the new biotechnology", in which they express, "the need for greater regulation and control of genetic engineering" and advise, "a moratorium on deliberate release of genetically engineered organisms (GEOs)" (Egziabher et al. 1995).

This situation of ongoing scientific dispute raises pressing questions in relation to the function that scientific knowledge may have in political decision-making. Controversy in applied science may enable policy-makers to choose between opposing scientific positions and thereby 'scientifically' legitimate their own preferred line of policy. This gives a rather ironic role to scientific expertise in the larger policy process. Scientific controversies leave room for policy-makers to choose the assessment of their liking. This undermines the role of science in applied contexts. Therefore, it is important to make an effort to analyse and evaluate scientific controversies in order to find out if there are no scientific reasons to prefer one position over another.

1.2.1 The terms of debate

The literature on the biosafety of transgenic organisms features a kaleidoscopic terminology. This mirrors the existence of diverting positions in the debate: our choice of words will reflect our ideas. The active role of language in our thinking has been characterized very aptly by Ludwig Wittgenstein in the observation that language, "... is the vehicle of thinking" ("... ist das Vehikel des Denkens" – Wittgenstein 1971: 168). Postman (1989) has extended this perspective by citing Wittgenstein rather liberally as saying: "Language is not just the vehicle of our thinking, it is also the driver" (Postman 1989: 27). Although there can be debate about the question in how far this extension does justice to Wittgenstein, the expression does shed a useful light on the driving forces of thought processes and discussions.

Abbreviations such as GEO, GEM, GMO, LMO ("living modified organism"), GMM, GEMMO, ONT ("organism with novel traits", cf. Doyle and Persley 1996: 30), have all been used to refer to transgenic (micro)organisms. The diverse uses of 'genetically modified organism', 'genetically engineered organism', 'living modified organism', etc., implicitly reflect a view on biosafety issues. Throughout this study, the construction 'genetically engineered organism' will be used because it makes it more clear that the impact of the changes made may not be known as well as the terms 'genetically modified' and 'living modified' seem to suggest (cf. Williamson 1992: 4). Thus, even a seemingly innocent aspect of the terminology in which one thinks about the larger problem may reflect a more serious problem of assessment. In the present study, 'GEO' refers to genetically engineered organism and 'GEM' refers to genetically engineered microorganism.

Those delving into the literature on the possible risks of modern biotechnology will also come across a host of terms and concepts with a more or less clear meaning, such as: 'acceptability', 'plausibility', 'familiarity', 'precision', 'analogy', 'precaution', 'uncertainty', 'sustainability', 'similarity', 'evidence', etc. Where they occur in this study, the chosen interpretation of those terms will be clarified. Following the cited interpretation of Wittgenstein, those terms may be studied as doing more than just articulate the debate, they also give direction to it. Throughout this study it will be argued that unclarity about the *definition of terms* is in large part constitutive of controversies and generally gets much less explicit attention in the biosafety debate than might be expected from a scientifically informed discussion.

One general distinction in terms that I use is between *hazard identification* and *risk analysis*. A 'risk' expresses the *chance* that some unwanted impact will occur, while a 'hazard' represents the unwanted impact itself (for more detail, see Section 2.2). The larger process of hazard identification and risk analysis, I refer to as *biosafety assessment*: "... a systematic means of developing a scientific basis for regulatory decision making" (Barnthouse 1992: 1751). The terms 'biosafety', 'hazard', 'risk' and 'harm' will often carry more meaning than can be articulated in scientific terms: "Neither risk nor detriment can be specified solely on the basis of scientific and engineering information. The selection of definitions necessarily involves social judgments about the factors to be included and the weights to be applied to them" (Dunster 1994: 139).

Another important element of unravelling the biosafety debate is to distinguish the general role of the actors involved in the controversies and their *authoritative legitimation*. For the purpose of this study, we must at least distinguish between *politicians, policy-makers, administrators, scientific experts,* and – last but not least – *lay people, consumers, citizens, 'the' public* (cf. Hill and Michael 1998).

For the purpose of understanding the character of biosafety controversies, an attempt must be made to distinguish the respective roles of these participants in the debate and their respective basis of legitimation. A claim on biosafety made by a politician, for example, will require a different basis of legitimation than does a claim on biosafety made by a scientific expert. Not making this (and other) distinction(s) will affect the argumentative transparency of the debate and thereby effectively obstruct a satisfactory resolution of ongoing controversies.

The terminology used by the participants in the debate is not neutral for the outcome. At the same time it must be noted that the terms of debate are often confused and seldom made explicit. This is a crucial obstacle to a proper understanding and to making the controversies fruitful for society. Confusion may keep controversy alive for no practical purpose. It is part of what Regal has called, "cryptic philosophy and ideology in the 'science' of risk assessment" (Regal 1996). This puts a burden on the debate, which merits the concern of scientists as well as of politicians.

Understanding a debate requires us to come to terms with its vocabulary. This requires the development of a practical way to explicate the terms of the biosafety debate.

1.2.2 Promises and perils of genetic engineering

The first genetically engineered plants were produced in 1982; by 1993 about 1200 plants were used in field releases; in July 1994 the 'flavr savr tomato' (genetically engineered for longer shelf-life) received a market license in the USA.

Early biotechnology companies were received with great enthusiasm in the financial world. By now, expectations have become more realistic. In 1998, the biotech companies of Europe and the United States were still facing annual financial losses that accumulated to over six billion US dollars (Pols 1998). However, the promise prevails and the industry is eagerly waiting for its first blockbuster to make the promise pay off (cf. Melcher and Barrett 1999).

Genetic engineering does not enable us to create just anything that can be designed in the lab. As Davis (1987) has phrased it: "... we must recognize severe limits to the power of molecular genetics to remake the living world. We can indeed recombine DNA at will, yet we do not have equal power to ensure survival of any recombinant: the requirements for a balanced genome determine what is viable at all, and then what can survive selection in nature" (Davis 1987: 1334). On the other hand, this kind of biological realism does not take away the fact that some expect genetic engineering in agriculture to be the key to such desirable objectives as *feeding the world population* and *preserving biodiversity* (cf. Altmann 1995).

Possible applications are numerous. Genetic engineering may be aimed at such diverse objectives as: disease control; enhanced plant growth; nitrogen fixation; crop protection against herbicides, insects, drought, frost, floods; biodegradation of toxic wastes; production of medicines. For example, crops such as corn, cotton, or potato can be genetically engineered to contain genes from the bacterium *Bacillus thuringiensis* (Bt), thereby enabling those crops to kill insects which feed on them. It is easy to be an enthusiast for the promises of genetic engineering: a crop producing its own insecticide certainly seems a desirable agricultural asset.

But is it a rose without thorns? Some say it is not. The *Union of Concerned Scientists* (UCS 1995) has expressed concern about possible perils: "Because the Bt toxin is produced in the tissues of the plants, it is protected fom the environment and does not break down. Under these conditions, pests feeding on the crops are exposed to high levels of toxin continuously over the life cycle of the crops. No better recipe could be found for eliciting resistance [of insects to the Bt toxin]. In fact, it is now widely agreed that in time the use of Bt-containing crops will lead to the development of resistant populations of pests" (UCS 1995: 15). This example shows how biosafety assessment must be part of a *cost-benefit analysis* for balancing the promises and perils of genetic engineering in agriculture.

Other fascinating promises of genetic engineering are: "a gene from the arctic winter flounder, whose liver proteins lower the freezing temperature of blood, into Atlantic salmon, thus enabling the salmon to survive in frigid waters they would not otherwise inhabit" (PEER 1995: 1); bananas producing medicine (Kiernan 1996); cells from genetically modified tilapia producing insulin in young diabetics (MacKenzie 1996); a strain of corn genetically engineered to secrete human antibodies (Gibbs 1997: 23). There can hardly be debate about the impressiveness of these possibilities.

Possible repercussions are also impressive: disruption of ecosystems by replacement of a naturally occurring species (Colwell *et al.* 1987: 22); the emergence of new viruses leading to new plant diseases (Schmidt 1995: 25); herbicide-resistant weeds; insecticide-resistant insects; transferred food allergens (Nestle 1996; Coghlan 1996).

To imagine the possible perils is almost as fascinating as imagining the possible promises: "There are no absolute rules governing the negative and positive aspects of a GEO release; the impact aspects cover immediate and long-term environmental effects. For genetically engineered plants, impact aspects could include: increased allergenicity of crop plants receiving genes from another plant species; the loss of genetic diversity through the widespread use of modified crops; the potential poisoning of native nontarget animals, fungi or bacteria by toxins expressed by engineered pest-resistant crops; and the unwanted vertical or horizontal transfer of genetic information (*e.g.*, antibiotic and herbicide resistance) from crop plants to weeds or bacteria" (Käppeli and Auberson 1997: 344).

The focus of the present study is on the *biological effects*, specifically. In the larger cost-benefit analysis, however, social-economic effects must also be considered. Small-scale farmers, for example, may become more dependent of biotechnology companies and possibly more susceptible to technological failures. In practice, the two types of impact are intimately related: when unwanted biological effects occur, this may have severe social-economic consequences as well (cf. van den Daele 1992: 327; Bunders and Radder 1995).

Another relationship between social-economic costs and biosafety assessment may be no less relevant: what if investments are postponed or cancelled on the basis of negative biosafety assessments that later turn out to have been misguided? The cost will then be in a *loss of possible benefit*. These high social and economic stakes only add to the urgency of making scientifically adequate assessments of the possible impact on our environment and our health of applying the products of recombinant DNA technology.

— 1.3 Science in biosafety assessment —

Given the new technology of genetic engineering and given the wish to assess the biosafety of applying this technology, the following question must be addressed: *what role can science play in the process of biosafety assessment?* This question will be central in this section; in the next section I will deal with the question of interpreting the occurrence of *controversy* in science for biosafety assessment.

A prerequisite for a reliable biosafety assessment is to develop an adequate account of what are the *boundaries* of the problem under consideration. These boundaries may not be immediately apparent. What may look like an important part of the problem may disappear from concern at closer inspection, and what remained unnoticed as a possible source of concern at first may turn out to be our biggest worry in the final analysis. Understanding the problem at hand is a process in its own right. When it comes to the assessment of highly advanced technologies such as genetic engineering, scientific understanding is a prerequisite to arrive at an adequate *problem definition.*

1.3.1 'Sensory organs' of science

The history of technology development has a counterpart in the history of technology assessment. Consider a historical example of the relationship between new technologies and new questions. A British commission assessing the possible effects of automobiles in 1908 thought the most serious problem of this new technology would be the *dust thrown up from untarred roads* (Collingridge 1980: 16). At the time, *dust* was a conceivable problem and thus a matter that raised critical questions. Accordingly, this technology assessment *avant la lettre* focused on the possible occurrence of a recognized environmental problem.

In his book *Risk Society* (1992), Beck has argued that modern societies must rely more and more on the interpretative work of the sciences to make the impacts of new technologies recognizable. In his view, "[t]he focus is more and more on hazards which are neither visible nor perceptible to the victims; hazards that in some cases may not even take effect within the lifespan of those affected; hazards in any case that require the 'sensory organs' of science – theories, experiments, measuring instruments – in order to become visible or interpretable as hazards at all" (Beck 1992: 27). If we consider the relevance of empirical observations and their scientific interpretation as a basis for biosafety assessment, it is indeed quite appropriate to think of the sciences as our 'sensory organs'. Without them we will not be able to *see* or *foresee* possible impacts of genetic engineering. Or, in other words, without the 'sensory organs' of science we will fail to arrive at a proper 'problem definition' for the purpose of biosafety assessment.

The new threats to the natural environment that are characteristic of our industrialized society share a certain 'invisibility'. To detect them we need scientific experiments. Without science, we would have no means to even be aware of their danger. Examples are *radioactivity, low-level toxic wastes, carcinogenic substances, deterioriation of the ozone layer, global warming*, etc. The risks they pose to our lives and our environment would be *invisible* to us without the help of scientific methods to detect them. In relation to the biosafety of genetic engineering, the 'invisibility' of possible hazards is a consequence of scientific uncertainty as well as of the fact that most biosafety assessments in relation to genetic engineering are *prospective.* They refer to a possible situation in the future, which must be made 'visible' or conceivable by scientific speculation (cf. Krimsky 1996a).

What can we do to make the 'sensory organs' provided by the life sciences more reliable? As a source of inspiration for a strategy to improve our scientific 'sensory organs' for biosafety assessment, consider the following example from the context of assessing the impact of *chemicals* on humans and the environment. Colborn and her colleagues recently wrote a new chapter in the history of environmental impact assessment and called it: *Our Stolen Future* (1996). They let their story begin with a list of what they depict as separate 'omens' of trouble: seemingly unrelated observations of worrisome environmental impacts of which it was unclear what might be causing them. Could there be a relationship, Colborn queried, between such diverse observations as: Lake Ontario gull chicks of which eighty percent died before hatching due to grotesque deformities; Lake Apopka alligators of which sixty percent of the males had abnormally tiny penises, obstructing effective pairing; Mediterranean Sea dolphins with partially collapsed lungs, breathing difficulties and abnormal movement and behavior; average human male sperm counts dropping by almost fifty percent between 1938 and 1990? (Colborn *et al.* 1996: 1-10). Could there be a hitherto unrecognized common cause to such distinct environmental impacts?

Most people are probably aware that chemical laboratories may contain dangerous substances. But who would have thought that even the seemingly innocent plastic tubes that are routinely used in laboratory work could be among the hazardous agents? Colborn and her colleagues brought together the evidence that some of the *plastics* (as well as other synthetic chemicals) that are widely used around the world may in fact be biologically active and act as *pseudo-estrogens.* If we consider the receptor of the female hormone estrogen as a lock that can be opened with the key called 'estrogen', then we must conclude that this 'lock' is somewhat uncritical; it turns out that keys other than the one it was intended for may also open it.

Since the chemical impact assessment procedures employed to test the possible hazardous effects of synthetic chemicals were conventionally focused on particularly detecting possible *carcinogenic* effects, another biologically detrimental impact remained undetected: synthetic chemicals acting as *pseudo-estrogens.* It took the probing interpretation of Colborn and colleagues to finally 'see' the underlying mechanisms and thus recognize the relationship between the initially separate observations of severe environmental impacts – 'omens' of a stolen future. Their work opened our eyes to a serious environmental problem and further developed our *scientific sensory organs* (cf. Boyce 1997).

It took a zealous interpretative attempt to finally make sense of a number of worrisome observations. Colborn and her collaborators finally uncovered a major environmental impact of what had seemed to be innocent synthetic chemicals before. The Colborn study is an excellent illustration of a fundamental aspect of scientific research, especially in relation to environmental impact assessment: doing research and recognizing the proper research questions takes time, dedication, effort, funds, etc. This is an essential practical element in debates on the scientific basis of biosafety assessment also, as will be further discussed in subsequent chapters.

1.3.2 A scientific detective

The subtitle of the book by Colborn *et al.* (1996) is '*A scientific detective story*'. In his foreword to the book, the author of *Earth in the balance* (1992) and US vice president, Al Gore, does justice to the detective-like character of this type of investigative research by stressing that Colborn and colleagues finally managed "to ask the right questions" about possible environmental impacts. Our capacity to make possible hazards 'visible' depends on our capacity to recognize the right research questions. A reliable scientific methodology would help us in the process of detecting the relevant questions to ask in relation to a specific environmental impact assessment. There is no guaranteed method to generate the most relevant questions for any specific case, but we can look for ways to discriminate between more and less relevant questions.

In many respects, applied science is more like a 'detective story' than a matter of applying ready-made scientific theories and laws. Most people know the satisfaction of learning the cunning and surprising outcome of a detective story. Things are brought to a solution in an unexpected way; there is an element of surprise in learning about and understanding a 'structure of causality' that we did not see immediately from the given 'facts'. The 'story' has to be uncovered with the help of ingenuity. The same goes for understanding the possible avenues of unwanted biological effects in general.

Similar episodes may be found in the larger history of environmental impact assessment. An early example of a 'hazard detective' is Carson (1962), who was the first to develop the 'sensory organs' to 'hear' the *Silent Spring* which was the result of the bioaccumulation of DDT in the food chain.

Or, consider the scientific detective work done by Molina and colleagues, who made us recognize the detrimental effect of CFCs (chlorofluorocarbons) on the Earth's *ozone layer* (cf. Nemeck 1997). The chemical stability of CFCs was applauded by industry in the 1930s, leading to their application in refrigerators, spray cans, foams, etc. It took until 1974 to recognize that there was one thing that could break down CFCs: sunlight. Ultraviolet radiation in the stratosphere turned out to be sufficiently energetic to break apart CFC molecules. The highly reactive chlorine atoms released in this process subsequently act as a catalyst in the process of breaking down ozone. Without our scientific 'sensory organs' we would be unable to 'see' this detrimental impact on the ozone layer.

Once we have learned to see through the eyes of people like Colborn, Carson and Molina and have apprehended their more sophisticated interpretations of environmental effects, our observations will change face. The history of environmental impact assessment is a history of *asking new questions* and thereby seeing new impacts. Overlooking possible future impacts on our environment may have severe consequences.

In biosafety assessment, the quality of the scientific sensory organs depends on whether we find the right research questions for the detective job. Just like a *physician* will rely on the medical tradition of asking relevant questions to make a patient's anamnesis, or like a *mechanic* will use a list of questions to check the condition of a car, a *scientific detective* will look for those questions that are relevant for a particular research purpose. All these jobs depend on whether their practitioners know which questions to address. In biosafety assessment this indeed requires the cunning and persistence of a Sherlock Holmes.

— 1.4 Controversies in applied science —

In the previous section, the general problem of appreciating the role of science in the context of biosafety assessment was discussed. Now, the additional problem of interpreting the occurrence of *controversies* in applied science, creating an insecure basis for biosafety assessment and policy (cf. Cambrosio *et al.* 1992), is considered.

The persistence of scientific controversy over policy-relevant issues will force politicians and policy-makers to make decisions on the basis of contested science. In technologically advanced societies this is an important concern. Finding ways to take science seriously, even if it does not produce consensus, is a *conditio sine qua non* for democratic societies.

1.4.1 Endless technical debate?

A challenging view on *controversy in applied science* is given by Collingridge and Reeve (1986). In their book *Science speaks to power – the role of experts in policy making*, Collingridge and Reeve introduce a, "skeptical model of science for policy". It presents a picture of the structure of scientific expertise and the genesis of controversy. Their model is a reaction to what they call, "myths of science", which are popular and widespread. The *skeptical model* rejects the idealized picture of a scientific enterprise that can effectively protect itself from external influences. According to the popular "mythical model", science must and can maintain "autonomy", "disciplinarity" and an "appropriate level of criticism"; science will only thrive when these three specific conditions are fulfilled.

First, the "autonomy" of the research process must be secured. This means that the progress of research should not fall under pressure from direct social interests. The second condition for good scientific research in the "mythical model" of science is that the subject of research must be confined to the more or less protective boundaries of one scientific discipline. When this "disciplinarity" is violated, it results in conceptual confusion and a paralysing debate about research perspectives. Finally, it is a "myth of science" that the internal scientific critique will not become too strong. A "moderate level of criticism" is a prerequisite for the creative process; it safeguards possibly useful ideas from being prematurely abandoned (Collingridge and Reeve 1986).

In applied science these three conditions will never be realized. The turmoil of the real world forces science to develop under less ideal circumstances. That is the main reason for Collingridge and Reeve to develop their *skeptical model* of science for policy. This alternative model specifies that the appraisal of a scientific hypothesis will always be under the influence of the *use* it will have in practice. Thus, the social world enters the world of science and the ideal circumstances for the process of scientific research will be disturbed.

The degree to which a scientific theory will be disturbed by practice is greater to the extent that the "error cost" of a conjecture will be higher. *Error cost* is defined by Collingridge and Reeve as the cost that a decision will have if it is made on the basis of a conjecture that turns out to be false (Collingridge and Reeve 1986: 30). Within the relatively protective boundaries of science this error cost will usually not be very high. The cost of an error within the scientific context will be mostly limited to loss of time, money and prestige of the involved researchers. When a conjecture is applied in a practical context, however, the costs can become much higher. Therefore, the adequacy of a conjecture is much more critical when science becomes relevant for policy. In such a situation science will fall under "stress" (cf. Collingridge and Earthy 1990).

In brief, the *skeptical model* consists of the following relationships: (1) Whenever scientific input to an important political controversy is wanted, scientific research will lose the "autonomy" that is an idealized prerequisite for research of good quality. (2) Whenever science is asked to evaluate practical problems, the relevant issues will usually transcend the boundaries of one scientific discipline. This will result in a violation of "disciplinarity". (3) Since a false conjecture can result in high costs when it is practically applied, the usually temperate level of criticism will be replaced by a "heightened or unreasonable level of criticism". (4) All this together will result in a situation in which no consensus will be reached, but instead, what Collingridge and Reeve call, an "endless technical debate" will prevail.

The irony of this perspective on applied science is that the actual influence of scientific research in matters which are important to society will be negligible according to Collingridge and Reeve, since no consensus will be reached. Consensus becomes impossible under the influence of disturbances that occur whenever science becomes relevant for policy. Collingridge and Reeve conclude from this that science will only have a real influence on policy, when it is concerned with matters of relatively minor political importance. In those cases, the scientific research process will not (or much less) be disturbed and consensus will remain a real possibility. In other cases, science will merely play an "ironical role". Its significance for policy will then only be in the fact that the rivalling scientific hypotheses keep one another in balance and therefore neither can have a decisive impact on policy. This leaves open the opportunity for policy-makers to create compromises.

What Collingridge and Reeve describe in their model of scientific expertise could perhaps, in caricature, be seen as 'the manufacture of controversy': "Autonomy is, of course, reduced when research is directed to questions posed by policymakers outside the scientific arena, producing a loss in the quality of the research, for such is the price of knowledge which is useful beyond the laboratory" (Collingridge and Reeve 1986: 30).

However, by recognizing a "loss of quality" Collingridge and Reeve implicitly suggest that it is possible to differentiate between higher and lower quality of scientific research. They nevertheless do not make an effort to explicate this possibility and thereby perhaps enable us to develop a *remedy* for this loss of scientific quality. Instead, Collingridge and Reeve conclude that we should seek compensation in the policy-making process for the quality that is lost in the scientific process.

In my view, the *skeptical model of science for policy* depicts a worrisome social process that should mobilize us to address the question: *what can be done to preserve the quality of scientific reasoning in a policy context?*

1.4.2 Loss of scientific quality in the biosafety debate

It is tempting to let oneself be desillusioned by this skeptical view on the role of applied science. Application of the skeptical model of Collingridge and Reeve to controversies in the biosafety assessment of genetic engineering indeed provides us with a challenging interpretation of the complex interaction between science and policy. To demonstrate just how persuading this model can be, I will now briefly sketch the biosafety controversies over genetic engineering along its lines. In the final analysis, however, I believe the model *does not* put us on the right track to an adequate understanding of scientific expertise in the risk society. Whereas Collingridge and Reeve take the "loss of quality" in a scientific debate to be a major (and identifiable) effect, they fail to look for a remedy within the context of science.

To see that the first disturbance of science for policy as predicted by the skeptical model, the *loss of autonomy*, is indeed a problem in the scientific research on the possible hazards of GEOs, it suffices to consider the roaring history of modern biotechnology. After the discovery of recombinant DNA techniques at the beginning of the seventies, promises have not only accumulated on the scientific side. For some time there has been a lively expectation that biotechnology would soon be a billion-dollar industry (cf. Kay 1993). This excitement has motivated many researchers to combine their scientific work with more commercial activities.

The following observation by Lewontin (1992) is perhaps typical for the situation: "No prominent molecular biologist of my acquaintance is without a financial stake in the biotechnology business. As a result, serious conflicts of interest have emerged in universities and in government service" (Lewontin 1992: 37). This is an indication of the fact that the so-called *error cost* (the cost that may result from false conjectures) of the biosafety assessment of GEOs may indeed be high and is not confined to biological hazards; it also involves the possibility of economical and social calamities, both for societies and involved individuals.

The second disturbance of science for policy that is foreseen by Collingridge and Reeve can also be found. To study possible hazards of GEO release we need to discriminate between different levels of biological organization. On the one hand there are *molecular* concerns about inserted or modified genes and their incorporation and expression in the host organism; on the other hand there are more *ecological* concerns about the traits of the transgenic organism in relation to its environment. Theoretically, these aspects can be distinguished, but in their application they are strongly related and intertwined. Integration of insights from different scientific disciplines (such as molecular biology, genetics, evolutionary biology and ecology) will be required to identify possible hazards that may result from the release of a GEO.

These different scientific disciplines not only focus on different research objects, but also work from diverging research perspectives. Therefore, it is not surprising that the need for *interdisciplinarity* does not develop without confusion over concepts and questions (cf. Krimsky 1991). A real-world problem will not restrict itself to the disciplinary boundaries of a scientific field. Thus, to get a grasp on the problem, scientific researchers are challenged to exchange views and positions with colleagues from other disciplines. This may easily lead to conceptual confusion and controversy. In 'pure' science this may be not so much of a problem, but in applied science this means that differences of opinion about an adequate interpretation of the research problem may accumulate.

The third disturbance of the skeptical model can also be recognized in the biosafety debate. An unreasonable level of criticism was made public by an investigation into the academic affairs at three major conferences about the biosafety assessment of GEOs by Wright (1986a). Apart from the fact that there had been selective publicity (which obstructs regular academic openness), the debates during the conferences were steered towards certain conclusions: "..., reservations about claims for decreased hazard were for the most part organized out of consideration, whereas claims that certain types of risk were minimal were organized in" (Wright 1986a: 598). This strategic procedure effectively led to the exclusion of critical positions from the debate. In the view of Wright, "..., as soon as genetic engineering was seen as a promising investment prospect, a fundamental deviation from traditional scientific practice occurred and a movement toward a corporate standard took place. The dawn of synthetic biology coincided with the emergence of a new ethos, one dominated by concerns external to science" (Wright 1986b: 336).

Collingridge and Reeve take the "loss of quality" that is effected by the process of consulting science for policy to be unavoidable. Considering their skeptical outlook on the general structure of scientific controversy this is understandable. In my view, however, the skeptical image of science is not inescapable. We are not compelled to take the "loss of quality" in science for policy as merely another empirical fact to cope with.

In the next section, I will present the preliminary outline of a strategy to confront this loss of scientific quality head on. Politicians get too much liberty of 'shopping' for expertise, because the notion of 'quality' in science has been lost in science studies of controversies (cf. Rip 1992; Segerstråle 1990, 1992, 1993). In Yearley's assessment of the role of science in environmental decision-making, "[t]he science studies view of knowledge and expertise allows us to appreciate the reasons the green movement has often found scientific knowledge an untrustworthy ally" (Yearley 1995: 477). Sociology of science has generated a number of such rather skeptical approaches to controversies in science. Most science studies look *outside of science* for compensations of the diverse forms of "loss of quality" they detect in science for policy (cf. Jasanoff *et al.* 1995; Nelkin 1992).

— 1.5 A methodological analysis of controversy —

From my acquaintance with the literature on biosafety assessment, I would say that virtually all arguments required to counter-balance the "loss of quality" that we can find along the lines sketched by Collingridge and Reeve, are already available in the debate. However, to appreciate the full spectrum of claims and arguments in the biosafety debate, we need a practical analysis that helps us weigh the opposing arguments and claims against each other. Such a tool may help us restore at least some of the scientific quality that may get lost in the heat of debate.

1.5.1 Contested plausibility of biosafety claims

Loss of quality in science can have far-reaching consequences. Von Schomberg (1997) has argued that science is in a stalemate due to controversial empirical claims and thus looses its functional authority. In his view, the only way to interpret what he calls an "epistemic debate" is by understanding the opposing biosafety claims as competing for "plausibility". The relationship between science and the risk society can be seen as a serious dilemma because of the paralysing effect of controversies on policy: "In the policy making process a contradiction arises: an appeal to science seems necessary (because of the complexity of the issue), but is not possible (since there is a controversy) and what is impossible (an appeal to a source which can provide authoritative data) becomes necessary" (von Schomberg 1993: 18). Our attempt to improve biosafety *policy* must begin with an attempt to develop an adequate image of biosafety *science.*

Von Schomberg (1997) gives a number of examples of contested biosafety assessments which he interprets as inescapable controversies between claims for plausibility. In his analysis, conflicting positions in biosafety controversies can be analysed as competing (clusters of) plausible claims (von Schomberg 1997: 78, 84). According to Von Schomberg, the opposing claims are indeed inconsistent, but the respective plausibility of the conflicting claims must still be granted, because the respective claims are based on reasonable arguments. In an "epistemic discussion", "we are forced to acknowledge that there is a lack of knowledge" (von Schomberg 1997: 84). I find this reconstruction unsatisfactory from a scientific point of view.

In my view, the term 'plausibility' is not clear enough in this context. With a more precise analysis we can demonstrate that at least part of these controversies can be analysed and evaluated with the help of methodological scrutiny.

Von Schomberg's argument that there is no way now to decide controversies, can be understood at least in part as a consequence of the "loss of quality" that was uncovered by Collingridge and Reeve (1986). My preferred answer to this problem is: let us try to restore some of the scientific quality and try to open up the scientific blackboxes that are interpreted by using the label 'plausibility'. Before asking *politicians* to make decisions, we must ask *scientists* to clarify the methodological status of their claims.

Alexander (1985) has also characterized the biosafety debate in terms of more or less 'plausible' claims: "The paucity of data did not restrain the spokesmen for either side. As capable and effective proponents of their positions, they gathered what facts there were, stitched these together with highly plausible speculation, and presented well-reasoned arguments – arguments that usually convinced those who needed no convincing and those who (often because of idiosyncracies of presentations by the news media) were exposed to only one side of the issue" (Alexander 1985: 58). Such controversial 'plausible' biosafety claims and one-sided arguments can be found not just in local newspapers, but in respected scientific journals such as *Science* and *Nature* as well (cf. Chapter 4).

To remedy a scientific "loss of quality", we must develop a framework for evaluating what Alexander has called "plausible speculation" and "well-reasoned arguments" at the core of biosafety controversies. How plausible is the speculation and how well-reasoned are the arguments in individual cases? By itself, the concept of 'plausibility' does not give us a tool for analysis, nor for evaluating contested claims. It does not help us to understand the role of scientific controversies in biosafety assessment.

Some have argued that the biosafety debate is burdened by irrational elements and that part of the arguments and claims that make up the controversy follow, "a patently absurd line of reasoning" (Kareiva and Parker 1994: 8). If we can find a way to assess the plausibility of individual speculations, it may help us find rational ways of dissolving (part of) the controversial issues. Others have dismissed references to possible hazards of genetic engineering as "myths" (Jukes 1988; Miller 1997). This leaves us with the challenge of recognizing and weeding out the "myths" and "absurdities" in scientific reasoning in the context of biosafety assessment.

Even if most "myths" and "absurdities" have already been shown to be misguided in the available critical literature on GEO biosafety, this has not led to a resolution of the debate. As Ho (1997) has argued: "One by one, those assumptions on which geneticists and regulatory committees have based their assessment of genetically engineered products to be 'safe' have fallen by the wayside. But there is still no indication that the new findings are being taken on board.

On the contrary, regulatory bodies have succumbed to pressure from the industry to relax the already inadequate regulations" (Ho 1997: 5).

Throughout the literature on the biosafety assessment of genetic engineering, virtually all the relevant claims and arguments have been both defended and criticized. In my view, this shows us that the biosafety debate is in need of a more basic interpretative framework to be able to analyse the contested claims and to evaluate the persisting controversies.

1.5.2 Claims and research questions: problem definitions

A major prerequisite of a pragmatic analysis of applied science is that the practical problem for which a solution is sought has methodological priority over the available theories that pertain to the larger field of research. In my view, there is no more practical way to define a research problem than in terms of the *research questions* that need to be addressed to understand the problem. Thus, I opt for an operationalization of the notion of a 'problem definition' in terms of *the relevant questions that must be addressed to reach a sufficient grasp of the problem, given a specified purpose of inquiry*. Such lists of questions can be seen as 'windows of concern' through which a problem is defined and interpreted. This approach implies that we may not have the answer to some of those questions.

For example, when Colborn and her colleagues set out to find the cause of a host of environmental 'omens', they finally came up with a new and extended *problem definition* for the practical purpose of assessing the environmental safety of synthetic chemicals. Until then, the problem had been defined in terms of one main question: 'Does a specific synthetic chemical have a carcinogenic effect?' Now, thanks to the scientific detective work by Colborn *et al.* (1996), we have a new problem definition as the basis of the safety assessment of synthetic chemicals. Apart from questions about possible *carcinogenic* effects, we now understand or 'see' that we must also ask questions about possible *pseudo-estrogenic* effects of synthetic chemicals to get a fuller grasp of the problem of chemical impact assessment. Leaving out the latter questions from a safety assessment would now amount to a wanton over-simplification of the problem at hand. The best recipe for future trouble is to start off with an over-simplified view of a challenge or, in other words, an oversimplified *problem definition* (see Figure 1.2).

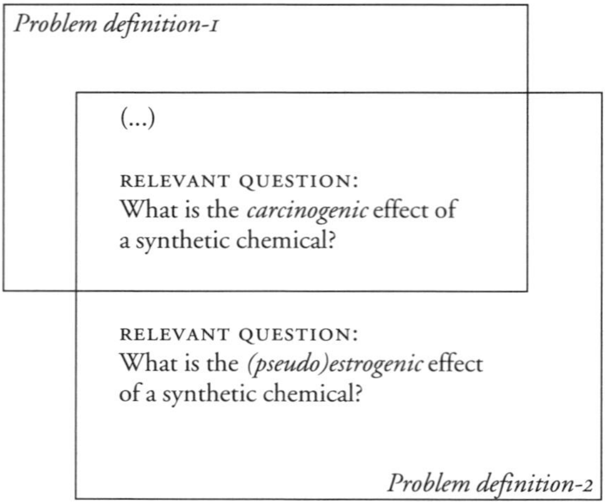
Problem definition-1
(...)
RELEVANT QUESTION:
What is the carcinogenic effect of a synthetic chemical?
RELEVANT QUESTION:
What is the (pseudo)estrogenic effect of a synthetic chemical?
Problem definition-2

FIGURE 1.2: Schematized representation of two alternative problem definitions or 'windows of concern' in relation to the biosafety assessment of the use of synthetic chemicals. Whereas at first the 'problem' was more or less routinely considered to be possible *carcinogenicity*, thanks to the 'detective work' by Colborn *et al.* (1996) the concern for possible *(pseudo)estrogenic effects* was added to the previous problem definition and thus to the research agenda for biosafety assessment of synthetic chemicals. A scientific claim on biosafety or health effects of a synthetic chemical based on *problem definition-1* will not suffice to establish the 'safety' of its use since the possibility of (pseudo)estrogenic effects was not taken into concern.

Part of the ongoing momentum of an "endless technical debate" is caused by the absence of an adequate problem definition. A debate about a proper definition of the problem in terms of the relevant research questions is less susceptible to a "loss of quality" than a biosafety debate in terms of competing claims for "plausibility". Before we can compare and evaluate the plausibility of claims, we must find out whether those claims are meant to address the same research questions. If it turns out they do not, then it is not possible to assess their comparative 'plausibility' in the given context. Many persistent controversies are essentially caused by hazy problem definitions. Controversies which arise over divergent problem definitions without being recognized as such, I call *artificial controversies.* Controversies which rise over the issue whether specific research questions should be included or excluded from the operationalization of a problem definition, I call *fundamental controversies.*

In my view, the *plausibility* of scientific claims in relation to a specific research problem is methodologically dependent on the questions upon which they are based. If we do not know which are the relevant questions, then a claim cannot be called 'plausible' for lack of reference. Thus, in claims for plausibility a claim for knowing the relevant research questions is implied. In practical contexts it may be difficult to assess the *plausibility of a claim* (even if we can agree about the relevant questions at issue). A more practical common ground for resolving an artificial controversy may be found in a focus on the *relevance of research questions* for the purpose defined (in this case biosafety assessment of genetic engineering).

The apparatus that enables us to decide whether a specific controversy is artificial or fundamental is the tool box of *general methodology* (cf. van der Steen 1993). The focus of general methodology is on the conceptual, interpretative and argumentative prudence that is a prerequisite of valid inferences. One example of it is the demand that one should avoid *ambiguity* of the concepts that are used in any theory.

How can this tool box of general methodology help us discriminate between artificial and fundamental controversies? I call those controversies 'artificial' for which it can be demonstrated that the conflicting claims address different research questions. This can be the result of ambiguity in terms and other conceptual unclarities. The nucleus of a debate is always a *problem.* A prior condition for a fundamental controversy to arise is that both antagonistic viewpoints address the *same problem,* viz. the issue of controversy. All this does not imply that an artificial controversy will necessarily generate less 'heat' than one that is fundamental. The concept of an artificial controversy is a heuristic tool; it is a searching device that would obviously be superfluous if all 'heated debates' were always fundamental controversies.

An important methodological background for assessing the relevance of research questions for particular research purposes, is the available background theory in a scientific field. For the purpose of this analysis, a problem is that theories may be involved in controversies themselves. This only duplicates the problem of evaluation. Therefore, I prefer an analysis of applied science that concentrates on the character of a research problem and on the way in which that problem is dealt with in the form of specified research questions. Of course, different background theories may be a source of relevant questions for different problem definitions. *General methodology* can be seen as the systematic research of our problem definitions in terms of relevant questions.

The central issue to be addressed is whether or not the questions that are explicitly or implicitly raised by the parties are adequate to deal with the issue of dispute. A problem is defined by the questions that we think relevant to ask about it. In cases of scientific conflict, this boils down to the question whether or not the theoretical background from which an expert is developing his or her arguments is suitably equipped to bear upon the problem at hand. Only in so far as a specific approach or theoretical background is arguably *useful* in relation to the problem that is the focus of disagreement, can it be part of a fundamental controversy.

The only way to check the *usefulness* of an approach with regard to a specific problem is to examine the relationship between research questions and the problem at hand. Only by conducting research about the exact meaning of concepts in different contexts, the generality of claims that experts put forward, the more and the less relevant aspects of a research problem, and scrutiny of all other aspects of reasoning that can go wrong; only by this kind of methodological research can it be decided whether or not a specific approach is useful to deal with a specific problem.

This does not imply that any problem will always have only one useful approach. In cases where two (or more) useful approaches are in disagreement we are facing a fundamental controversy. I argue that in some cases we can diagnose controversies as being *artificial* rather than *fundamental*, by demonstrating that one (or more) of the proposed approaches does not adequately address the problem of concern. Such a diagnosis leads to the remedy of *dis*qualifying those approaches that turn out *not to be useful in view of the specified research purpose* (cf. Kooijman 1993: 8; van Dommelen 1996a).

The relative usefulness of a problem definition in view of a specified research purpose will be discussed in further detail in Chapter 2. The *usefulness of a problem definition* for a specific purpose will be operationalized in terms of the *relevance of research questions* in relation to that purpose.

— 1.6 Biosafety analogies and *artificial* controversy —

In this section, I give a preliminary demonstration of the clarifying potential of the practical analysis introduced above, by applying it to a long-standing controversy in the biosafety debate. More detailed analyses and more specific demonstrations will be presented in subsequent chapters.

A recurring theme throughout biosafety controversies and throughout this study is the quest for *sufficient knowledge* for the purpose of assessing the biosafety of genetic engineering. One type of 'vehicle of thinking' that is frequently used to claim sufficient knowledge is the argument by analogy. The use of *analogies* to support a line of argument pro or con the supposed safety of specific deliberate environmental releases of genetically engineered organisms is widespread in the literature on GEO biosafety assessment. So much so that analogies have come to play an important role in the debate as a whole. Below, some central examples are cited and it will be argued that such analogies only *seem* to give an answer to the question of sufficient knowledge and scientific uncertainty.

The reason for the use of analogies being an inadequate solution to the problem of sufficient knowledge here is that analogies are inadequate means of problem definition for the purpose of biosafety assessment. Reasoning by analogy in this context implies that the relevance of specific research questions remains implicit; thereby enhancing the danger of comparing 'plausible' claims that do not address the same research questions. Thus, the use of analogies as tools for problem definition reflects a loss of quality in applied scientific research. They are an attempt to interpret a new problem through the 'sensory organs' of previous experiences. Common ground can only be found by explicating the implied concerns. Two main analogies can be disinguished as recurring elements in biosafety controversies over GEO release: the *domesticated species analogy* versus the *exotic species analogy*.

The *domesticated species* or *traditional breeding* analogy is often used by opponents of new regulation for recombinant DNA technology. At the basis of this general argument is the idea that the 'improvements' that are produced by recombinant DNA technology must be seen as merely the next evolutionary step in a long tradition of agriculture: "The degree of novelty of microorganisms or macroorganisms created by the new genetic engineering techniques has been widely exaggerated. A corn plant that has incorporated the gene for and synthesizes the *Bacillus thuringiensis* toxin is still, after all, a corn plant" (Young and Miller 1987: 23).

The general conclusion Young and Miller (1987) arrive at is: "Thus, a more applicable model for organisms modified by the techniques of new biotechnology is the selective breeding and testing of domesticated plants, animals and microbes, with which there is vast experience and which boasts an admirable safety record" (Young and Miller 1987: 26). The general basis of this biosafety claim is the assumption that 'domesticated species' are so specific for the controlled conditions of agriculture, that they will not be able to survive in natural situations. An example of this is Federoff's suggestion to use corn as a "model organism", which is unable to spread without human intervention (Fedoroff 1991).

Another group of experts in the biosafety debate on the deliberate release of GEOs relies on what is called the *exotic species* analogy. One of its advocates argues that, "[t]he introduced species model, or analogy, is nevertheless valid because the ability of an organism to become established and spread in a new environment is an ecological question regardless of the origin or nature of the "new" or "novel" aspects of that organism. In other words, the ecological processes that are called into play when an organism is moved from one place to another are the same whether that organism is moved from one continent to another, from one ecosystem to another on the same continent, from one environmental medium to another in the same ecosystem, or from a laboratory to an open field. (...) I suggest that it is fruitless to argue much longer over the validity of the exotic species model as an analogy for introduced recombinants" (Sharples 1987: 94-96; cf. Regal 1987).

Whereas the domestication analogy is usually promoted by participants in the debate who think the risks are (almost) negligible, the *exotic species analogy* is mostly put forward by experts who make it clear that they are not so sure about that. The basic analogy of introduced or exotic species is that experience has learned that they may become a nuisance or a pest by lack of natural competitors or predators. Well-known examples are the introduction of rabbits in Australia, and of the gypsy moth and chestnut blight in the United States (cf. Sharples 1983). Most of the cases in which the introduction was 'unsuccesful' and therefore harmless, are unkown to us. As a rough estimate a ten-ten rule has been suggested: "... 10% of species introduced become established and (...) 10% of those established becomes pests ..." (Williamson 1994: 75).

The pervasive role of analogies in debates about the biosafety of deliberate environmental release of GEOs, in my view represents what Collingridge and Reeve have called "a loss of quality" of applied science in a policy context. The analogies cover more than is warranted and could bring any debate out of balance. Reliance on the use of analogies in the biosafety debate has enhanced controversy and endless debate rather than resolved it. This is unsatisfactory from a scientific perspective and leaves room for policy-makers to rely on the 'expertise' that suits their purposes best.

1.6.1 Contested analogies reconsidered

The suggested analogies can be seen as reflections of different *problem definitions* for the purpose of assessing the biosafety of transgenic organisms. Experts that support the *domesticated species analogy* essentially claim: 'The relevant questions for the biosafety assessment of transgenic organisms are the same as the relevant questions that are now in use for the biosafety assessment of applying non-transgenic organisms'. Experts supporting the *exotic species analogy*, on the other hand, claim that the relevant questions for GEO biosafety assessment should include other concerns as well. In this reconstruction of these disputed analogies, the controversial scientific issue to resolve is: *which of the implied problem definitions would be more useful for the purpose of GEO biosafety assessment?*

From the perspective of a practical analysis, the main issue to evaluate in these controversial analogies is to find out which of the two implied problem definitions would be more useful for the purpose of biosafety assessment. And, if these 'competing' problem definitions are made explicit, the analogies themselves can be dismissed altogether. In this analysis, the main reason why the reliance on analogies tends to lead to endless controversy is that the associated problem definitions remain largely implicit and thereby unclear. This makes it impossible to decide which of the two contested analogies would be more 'adequate' or 'useful in view of the research purpose' in this context. The only feasible way to analyse and evaluate them is in terms of the implied relevant questions.

The general scheme of inference underlying the use of analogy can be expressed as follows: "... a,b,c,d all have the properties P and Q; a,b,c all have the property R; therefore d has the property R..." (Leatherdale 1974: 10). We can apply this scheme of inference to the contested analogies. When a participant in a discussion on the biosafety of GEOs claims that we can interpret their application by analogy with the extensive agricultural experience with domesticated species and traditional breeding, for example, the implied argument goes as follows: 'Domesticated organisms (a,b,c) and transgenic organisms (d) have several properties in common (P and Q); domesticated organisms (a,b,c) are safe in agricultural use (R); therefore transgenic organisms (d) will be safe in agricultural use (R).'

It is not very difficult to list several properties that domesticated and transgenic organisms have in common. But this does not, by itself, justify the claim that therefore transgenic organisms will be as safe as conventional organisms in agricultural applications. One problem is the implied generality of the claim. Maybe (probably) it is indeed true that many transgenic organisms are just as safe as conventional organisms (although it should be noted that conventional organisms have caused serious unwanted effects and are thus not always as 'safe' as the users of this analogy would like us to believe). But then we must still face the problem of deciding which, if any, of the transgenic organisms could be the exception to this 'rule' (if there is a rule here). The use of analogies in this context boils down to asking questions about a,b,c and making claims about d. To justify this, a specification is required of a,b,c, *and* d, or, in other words: the use of analogy must be specified in terms of the research questions that are considered relevant to address.

In this analysis, the sensible way to appreciate the use of analogies in the context of biosafety assessment is to interpret them as heuristic resources to arrive at relevant questions for the purpose of GEO biosafety assessment. From this perspective, evaluation of the two analogies in view of a useful problem definition should be focused less on their conflicting or *competing* elements and more on their *complementing* elements. One analogy may help to appreciate some possibly relevant questions and another analogy will do the same for other possibly relevant questions. Neither of the two separate analogies can give a sufficiently useful problem definition by itself; taken together and appreciated in their own right, they may help us arrive at a more useful problem definition than either of the two could have done individually (see Figure I.3).

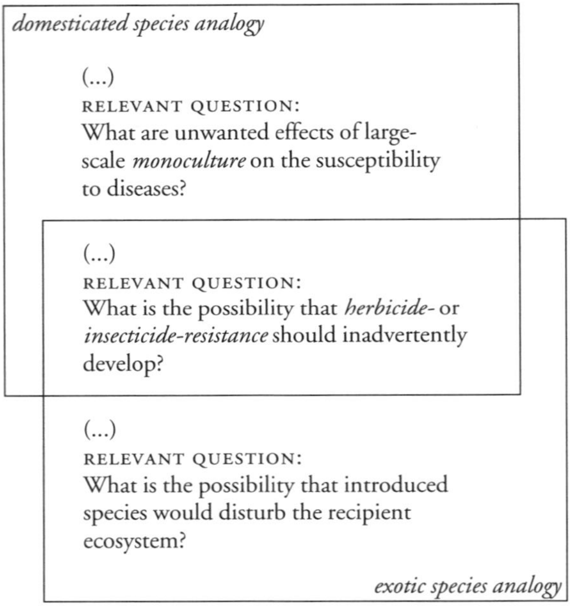

domesticated species analogy
(...)
RELEVANT QUESTION:
What are unwanted effects of large-scale monoculture on the susceptibility to diseases?
(...)
RELEVANT QUESTION:
What is the possibility that herbicide- or insecticide-resistance should inadvertently develop?
(...)
RELEVANT QUESTION:
What is the possibility that introduced species would disturb the recipient ecosystem?
exotic species analogy

FIGURE I.3: Schematized representation of two alternative problem definitions or 'windows of concern' in relation to the biosafety assessment of genetically engineered organisms. The representation shows that alternative analogies do not reflect concerns in a format of *either-or*, but rather in a format of *and-and*.

The analogy to the experiences with traditional breeding, for example, may help us appreciate the relevant questions of conventional agriculture (cf. Doyle 1986). The worrisome speed with which insects may develop resistance to insecticides or with which weeds may become resistant to herbicides, are just two of the serious unwanted effects from conventional agriculture that should make one concerned about asking the related relevant questions in GEO biosafety tests also. It should be kept in mind that conventional agriculture is by no means a 'model' for the absence of risks or of 'trouble-free' technology. Traditional breeding has led to serious problems such as species uniformity, pathogen invasions, pest resistance, reduced biodiversity, herbicide resistance (cf. Mantegazzini 1986: 76-80). Some have even presented genetic engineering as a possible 'cure' for the backlashes of traditional agriculture: "Transgenic plants and microorganisms can help to diminish the negative environmental effects of intensive agriculture" (Schell 1994: 18).

At the same time, the proposed analogy to intended or unintended introductions of exotic species can also help us arrive at relevant research questions for the purpose of GEO biosafety assessment. Introduced exotic species have caused havoc and serious unwanted effects all around the world and these experiences should not be overlooked as sources of possibly relevant questions for biosafety assessment either. This does not take away the problem that, by itself, this analogy is not specific enough to give a scientific basis to assess the possible risks of biotechnology either: "In its presently unrefined state, the non-indigenous species comparison would overestimate the risks of GEOs if it were (mis)applied to genetically disrupted, ecologically crippled GEOs, but in some cases of wild-type organisms with novel engineered traits, it could greatly underestimate the risks. Further analysis is urgently needed" (Regal 1993: 225).

Thus, in the practical methodological analysis as introduced in this chapter, the "endless" controversy over "plausible" analogies to be applied as models for GEO biosafety assessment is an *artificial* rather than a *fundamental* dispute because it evades the challenge of specifying the implied problem definitions in terms of the relevant research questions. A fundamental analysis in terms of the implied problem definitions shows us that the contested analogies should be evaluated as *complementing 'windows of concern' for GEO biosafety assessment* rather than as conflicting scientific positions.

The general approach to applied science in biosafety controversies presented in this introductory chapter will be further specified in Chapter 2 and further applied and demonstrated in subsequent chapters. The challenge for a methodological analysis of "endless technical debate" among scientific experts is to recognize 'artificial controversies' and translate them in the underlying 'fundamental controversies', thereby weeding out the spurious elements of a dispute.

CHAPTER 2

Evaluating Contested Claims on Hazard Identification

— 2.1 Testing for biosafety —

The most reliable way to find out whether the intentional release of a particular transgenic or genetically engineered organism (GEO) could lead to inadvertent consequences would be to just go ahead and try it. Most deliberately released GEOs would probably turn out to be quite harmless (cf. Regal 1996: 15). But, what if a small fraction of GEOs would lead to serious trouble? We would regret having applied those GEOs and would find ourselves wishing we had a test of some kind to find out which GEOs pose a threat to humans or the natural environment and which do not. What would a biosafety test like that look like? Experts can be found over the full spectrum of opinions, ranging from "no need for testing" to "a moratorium on testing for safety reasons". In my view, the fundamental issue in biosafety controversies over the deliberate release of GEOs is: what would be a sufficient test to assess the hazardous potential of a particular GEO?

A test is a tool for answering questions. A particular test will address some specific questions and neglect others. A lab test, for example, will not answer all questions that can be addressed by a field test. No single test can address all questions. Thus, to design a test for the hazardous potential of a GEO, we need to know what are the relevant questions to address for this research purpose. A particular biosafety test will be a reflection of a particular definition of the problem at hand. In this reconstruction, scientific disputes over biosafety assessment are essentially about the issue: what are the relevant research questions that must be addressed to identify the hazardous potential of a GEO? Biosafety controversies can be interpreted as disagreements between experts about what is considered to be a sufficient 'set of relevant questions' (for definition see below) for the purpose of GEO hazard identification.

All research depends upon the questions raised. To be able to evaluate some scientific claim, we must specify the research questions that were addressed before arriving at this claim. The specific list of relevant questions that a scientist considers necessary to address in a biosafety assessment is an articulation of the underlying (implicit or explicit) problem definition. For the purpose of this analysis, I introduce an analytic tool which I call a researcher's 'set of relevant questions'. The concept of a 'set of relevant questions' (for brevity: SRQ) is defined as 'a collection of research questions that a scientist considers relevant for the study of a specified research problem'. A researcher's SRQ for a specific research purpose is the most explicit formulation of a problem definition. This explicitness is a prerequisite for a scientific approach of a practical problem.

Different research problems will require different SRQs to live up to the respective research purposes. Scientists may (and often do) disagree about what is a sufficient or useful SRQ for a specific research purpose; this reflects disagreement about an adequate problem definition of the research subject. To compare the empirical status of contested claims, we must first ask whether they are meant to be an answer to the same research question(s). Discussions about what constitutes a sufficient test can be reconstructed in terms of the *relevance* of individual scientific questions in view of the research purpose.

In this chapter, a framework is presented for *analysis* of contested biosafety claims in terms of their underlying research questions and for *evaluation* of the relevance of particular questions on the basis of a *scientific burden of proof* for including or excluding specific research questions for the purpose of GEO hazard identification. In subsequent chapters, the analytical and evaluative potential of this approach will be demonstrated by applying it to long-standing biosafety controversies over the deliberate environmental release of transgenic organisms.

2.1.1 Choice of research questions

A controversy over an adequate problem definition as a basis for a biosafety test can be reconstructed as a contested choice between two alternative possible SRQs. For example, SRQ-1 may encompass Questions a, b, c, and d as relevant research questions, while SRQ-2 encompasses only a, b, and c. Competing SRQs can be 'compared' or evaluated by discussing the relevance of the individual questions that make up a larger SRQ. A methodological procedure to evaluate this relevance will be discussed in Sections 2.3 and 2.4. A preliminary demonstration of an analysis in terms of competing SRQs will be presented in Section 2.5.

As an example of how explicit questions make a test, consider the *Performance Standards for Safely Conducting Research with Genetically Modified Fish and Shellfish* developed by the US *Agricultural Biotechnology Research Advisory Committee* (ABRAC 1995a, 1995b). The core of this approach is formed by a collection of "flowcharts", which represent a trajectory of relevant questions for a specified research purpose in GEO biosafety assessment. The ABRAC presents flowcharts of questions as prerequisites for making claims on such concerns as: "Impact of deliberate gene changes", "Impact of interspecific hybridization", "Potential interference with natural reproduction", "Ecosystem effects", etc. (ABRAC 1995b).

The flowcharts of the ABRAC are exemplary in their attempt at explicitness. The resulting biosafety claim is an answer to the trail of prescribed questions and can thus be interpreted against the background of these specific questions addressed. The claim will only make sense if we see it against the background of the listed questions; the questions 'make' the claim (see Figure 2.1).

FLOWCHART QUESTIONS (ABRAC 1995b: II.A)

> QUESTION:
> "Does the GMO result from deliberate changes of genes?"
>
> QUESTION:
> "If containment is removed, does the GMO have direct access to (a) suitable natural ecosystem(s)?"
>
> QUESTION:
> "Is/are the accessible ecosystem(s) isolated from other aquatic ecosystems and of low enough concern that killing of all fish/shellfish in the event of a GMO escape would be possible and practical?"

FIGURE 2.1: Schematized representation of one trail of questions as part of a Flowchart for the purpose of "Safely conducting research with genetically modified fish and shellfish", as developed by ABRAC (1995a, 1995b). Answering 'Yes' to all three cited questions would give a green light to the proposed research.

To compare the empirical status of conflicting claims, we must compare the flowcharts or SRQs that produced them. Claims can only be evaluated against the background of the research questions to which they are an answer. In practice, the specific research questions at the basis of a biosafety claim are often left implicit in the debate. Reconstructing and comparing the underlying flowcharts or SRQs allows us to analyse and evaluate the claims that are based on them. As we shall see, there are different restrictions to our choice of questions. Practical as well as theoretical concerns will keep us from asking all possibly relevant questions. Different research purposes will require us to concentrate on different research questions. For example, a scientist who wants to predict tomorrow's *wheather* will have to address different research questions and apply a different SRQ than a scientist will need who wants to predict next centuries *climate.*

In terms of the proposed reconstruction, the flowcharts developed by ABRAC are reflections of different *sets of relevant questions* for the purpose of assessing the biosafety of transgenic organisms. Attempts to make the reseach questions explicit, which is constitutive of the ABRAC approach, are sorely missing in the more general controversy on the biosafety of genetic engineering. As a preliminary illustration of this omission, consider such "notification requirements" as listed in *Annex II* of EU Directive 90/220 (see also Section 5.5.1). The requirements listed there do no specify how the required information should be produced, leaving room for controversial interpretations and too much freedom to the notifier. The underlying problem is that it does not provide an explicit problem definition in the form of a sufficient SRQ. For the purpose of biosafety regulation, the specification of the relevance of research questions for biosafety tests is a primary requirement.

The concerns listed in *Annex II* are an important first step towards a scientific approach to biosafety assessment, but they leave open too much possibility of addressing those concerns in very different research styles. The only way to compare and evaluate the usefulness of standards for biosafety assessment is to articulate the concerns they cover in terms of explicit research questions to be addressed in a biosafety test. In this respect, the ABRAC flowcharts are an exemplary attempt to give biosafety testing a scientific basis.

The scientific issue to resolve in biosafety controversies is: which of the implied sets of relevant questions underlying opposing claims would be sufficient for the purpose of a GEO biosafety test? Fundamental controversies in biosafety assessment (as opposed to artificial controversies) should be about what constitutes an adequate biosafety test for a specific GEO. Such fundamental scientific discussions focus on adequate problem definitions in the form of the relevant research questions for the research purpose.

— 2.2 Risk analysis and hazard identification —

Quantitative *risk analysis* presupposes qualitative *hazard identification.* Risk can be defined as the product of hazard and likelihood: Risk = Hazard x Likelihood. Alexander (1990) has characterized the difference as follows: "Risk analysis separates hazard from risk. (...) The hazards are the possible detrimental effects. In contrast, the risk may be 'one in five' or 'one-in-a-million', *i.e.*, the likelihood of harm. If exposure does not occur, there is no risk, regardless of the hazard. If there is no hazard, there is also no risk, regardless of exposure" (Alexander 1990: 121; cf. Barnes and Hulsman 1995: 278).

An alternative phrasing is suggested by Bergmans (1995): "Risk is usually defined as hazard times the frequency that a certain hazardous situation actually occurs" (Bergmans 1995: 23; cf. NSNE 1997: 4). Use of the term 'frequency' in this context may serve as an illustration of the misunderstanding that may easily arise in the interpretation of 'likelihood' and 'probability', which are more concerned with *chance of occurrence* than with *regularity of occurrence* (= frequency). The main point to note here is that in too many discussions on biosafety the distinction between *risk* and *hazard* is not explicitly made (cf. Barnthouse 1992). This is an omission which may cause confusion, and therefore: "In order to evaluate a risk, the hazard associated with it must be clearly identified" (Bazin and Lynch 1994: xii).

Consider, as an example, the distinction between hazard and risk in relation to the 'millennium problem' discussed earlier. The 'hazard' that has been identified (albeit not early enough) is that computer systems will fail to operate as expected when their year counter jumps from 99 to 00. The 'risk' of this happening depends on many factors and can only be assessed for individual computer systems. However, given the identified hazard it would be unwise to speculate over a low risk or likelihood of occurrence.

Fundamental controversies in the biosafety debate are more often about 'hazard identification' than about 'risk analysis'. Where there is agreement about the existence of a possible hazard, there will usually also be agreed concern about the existence of a risk. Therefore, the focus of this study is on the scientific basis of *hazard identification*, rather than on the outcome of *risk analysis*.

Hazard identification is the attempt to recognize possible unwanted effects of some endeavor. When a particular GEO is applied in the field, what consequences may be expected? Before something can be said about a possible risk, the possible hazards must be identified. Hazard identification is the kernel of biosafety assessment, as the following may illustrate: "In most cases, the identification of any realistic hazard associated with a biotechnological application was sufficient grounds for stopping a project in its early stages, avoiding any risk" (Käppeli and Auberson 1997: 347).

Qualitative hazard identification is a prerequisite for *quantitative* risk analysis. Whatever is not included as a matter for concern in a hazard identification, will not be 'seen' or considered in a risk analysis. Therefore, a methodological priority for improving the process of GEO hazard identification is to develop a method that minimizes the possibility of *overlooking* relevant biological concerns. For that reason the present study focuses on the methodology of hazard identification.

In the hazard identification of an intentional or deliberate release of a GEO, biosafety assessors look for information about its possible environmental impact. The *purpose* of GEO hazard identification is to recognize possible unwanted effects, in so far as they exist. But what information exactly is needed to satisfy this purpose? The first thing to know about a specific hazard identification is: *which questions need to be addressed?*

Ongoing biosafety controversies such as introduced in Chapter 1 are essentially *disputes about what are relevant questions for GEO hazard identification.* Disputes about adequate SRQs are a prerequisite for appropriate hazard identification and biosafety assessment.

2.2.1 Identifying hazards of transgenic Klebsiella planticola

Empirical research cannot be trusted at face value. Consider an episode in the recent history of GEO hazard identification as an illustration of how important it is to recognize and articulate the supposedly relevant questions for a particular research purpose.

When biosafety assessors were asked to study the possible impact of genetically engineered *Klebsiella planticola* (a soil bacterium genetically engineered for an enhanced production of ethanol from agricultural residues in closed containers and to be spread onto the field as compost after ethanol production) on wheat plants, they found that this genetically engineered microorganism did not affect the development of wheat, a major agricultural crop plant.

These observations seemed reassuring enough, until an alternative interpretation of the empirical results was suggested. Holmes and Ingham (1994) raised the question whether it could make any difference for the interpretation of these observations that the test of the impact of *Klebsiella planticola* on wheat plants had been done in *sterile soils.* To answer this question, they set out to assess the impact of genetically engineered *Klebsiella planticola* on wheat plants *in natural soils* (Holmes and Ingham 1994, Holmes *et al.* 1999). The results of their tests led to a more qualified interpretation of the earlier tests in sterile soil.

Holmes and Ingham (1994) found that genetically engineered *Kl. planticola* affected the *rhizosphere* of soil microorganisms which in turn affected the wheat plants. In comparison to the parental strain, the genetically modified strain had a significant impact on soil foodweb organisms, such as nematode species feeding on bacteria and fungi, resulting in a decrease of plant growth, a fully unexpected but ecologically significant effect (cf. NSNE 1997). Inclusion of the role of soil microorganisms as a relevant concern in the biosafety test and in the interpretation of possible environmental impacts revealed limitations of the earlier observations. The earlier impact assessment had found "no effects" because it had been based on a too limited set of research questions. Recognition of the intermediary role of the rhizosphere in this particular case led to prevention of what could have become a severe environmental impact.

In terms of general methodology, what Holmes and Ingham did in their additional research was to introduce a new problem definition and thus to apply an alternative SRQ for the purpose of transgenic *Klebsiella planticola* hazard identification. The result was the detection of a hazard potential that had been left undetected before. The scientific 'detective work' (cf. Section 1.3.2) done by Holmes and Ingham demonstrates the importance of using an appropriate SRQ in view of the purpose of hazard identification. What they found is not a full-blown biological theory that can predict the effects of genetically engineered *Klebsiella planticola* on wheat plants. What they found is much more modest than that and at the same time the most important prerequisite for the purpose of biosafety assessment. They found that the *set of relevant questions* (SRQ) for recognizing the possible impact of *Klebsiella planticola* should be extended in comparison to the set of questions that was at the basis of earlier laboratory tests. By including the rhizosphere in their considerations, they essentially extended the list of *relevant questions* to ask about the possible hazards of genetically engineered *Klebsiella planticola.*

The respective experiments lead to opposing empirical claims about the biosafety of applying genetically engineered *Klebsiella planticola.* One claim does not recognize any hazard of this application, whereas the other claim does. To evaluate their respective trustworthiness for the purpose of biosafety assessment, we can analyse the opposing claims in terms of the respective SRQs underlying the respective claims. For this purpose we will focus on the *differences* between *Set of relevant questions-1* (underlying the claim that transgenic *Kl. planticola* poses no hazard) and *Set of relevant questions-2* (underlying the claim that transgenic *Kl. planticola* does pose a hazard). Evaluation of the two opposing biosafety claims boils down to the: question which of the two alternative SRQs underlying the two experiments producing the empirical results is more relevant or useful for it, purpose: SRQ-1 or SRQ-2?

The essential difference between SRQ-1 and SRQ-2 is that Holmes and Ingham decided to include the complex cluster of soil microorganisms that is referred to as the *rhizosphere* into consideration in their alternative experimental design. In SRQ-1 the rhizosphere is not considered, because the experiment based on SRQ-1 used sterile soils. Thus, the implied assumption of this experiment and its interpretation was that the rhizosphere was not a relevant concern for the purpose of transgenic *Kl. planticola* hazard identification and that the role of soil microorganisms need not be included in a SRQ for this purpose.

By including the rhizosphere in their assumed SRQ, Holmes and Ingham included new biological relations as possibly relevant concerns for transgenic

Kl. planticola hazard identification. In SRQ-1, the main relationship conceptualized as being of possible concern is the direct relation between genetically engineered *Klebsiella planticola* and wheat plants. The experiments based on SRQ-1 demonstrated that this direct (or isolated) relation does not pose a threat to the wheat plants. Apart from the direct relation between *Klebsiella* and the wheat plants, in SRQ-2 two other relations are also considered: the relation between genetically engineered *Kl. planticola* and the other soil organisms *and* the relation between the rhizosphere and the wheat plants. From the perspective of hazard identification, this turned out to be an essential difference (see Figure 2.2).

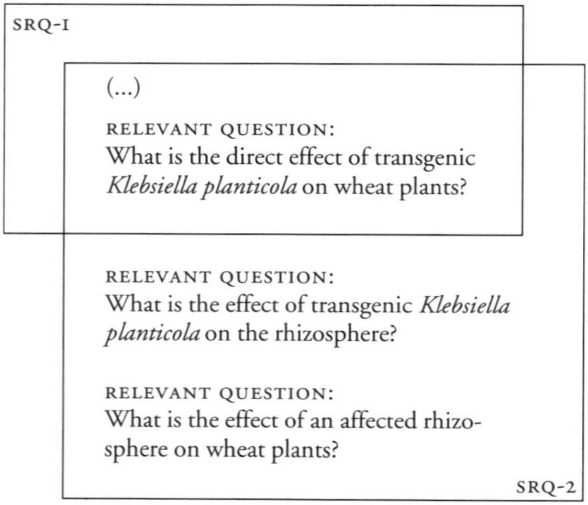
SRQ-1
(...)
RELEVANT QUESTION:
What is the direct effect of transgenic *Klebsiella planticola* on wheat plants?
RELEVANT QUESTION:
What is the effect of transgenic *Klebsiella planticola* on the rhizosphere?
RELEVANT QUESTION:
What is the effect of an affected rhizosphere on wheat plants?
SRQ-2

FIGURE 2.2: Schematized representation of two alternative SRQs (or 'windows of concern') for the purpose of assessing the biosafety of transgenic *Klebsiella planticola*. The questions in SRQ-1 have left undetected a detrimental effect that was detected thanks to SRQ-2.

The experiments based on SRQ-2 demonstrated that genetically engineered *Kl. planticola* affects the rhizosphere. Thus, in the scope of SRQ-2 an effect was included that could not be detected by SRQ-1: an impact of genetically engineered *Kl. planticola* on the rhizosphere. This impact subsequently changed the relationship between the rhizosphere and the wheat plants. Whereas the control rhizosphere provided an excellent basis for the wheat plants to grow, the rhizosphere affected by genetically engineered *Kl. planticola* led to disease and death of the wheat plants. SRQ-2 had enabled Holmes and Ingham to identify a hazard that was left unrecognized by SRQ-1. Their extended SRQ included questions about a possible impact of transgenic *Kl. planticola* on the rhizosphere and questions about a possible impact of a changed rhizosphere on wheat plants. *Asking the right questions* led to a more adequate GEO hazard identification. If a 'flowchart' for this research purpose (cf. Section 2.1.1) would be based on a choice between either of the two SRQs, the decision would have to be made in favour of SRQ-2.

Although I consider this illustration as representative for the larger challenge of biosafety assessment, it is a relatively simple example because of the present availability of empirical results. In biosafety assessment it is often impossible to perform such direct tests and thereby demonstrate the relevance of research questions. This makes it the more important to approach biotechnology hazard identification in terms of relevant *questions* rather than in terms of conflicting *answers*.

— 2.3 Science as the art of asking questions —

A famous episode in the history of science has been related by Carl Hempel (1966). He describes how Semmelweis, who was a physician in a Vienna hospital from 1844 to 1848, developed different hypotheses to understand what was going on in his clinic. Semmelweis' applied scientific work can be seen (and is in fact represented by Hempel) as the work of a detective who is hunting after a helpful interpretation of his observations (high instances of women who died after giving birth). Semmelweis had no ready-made theory waiting to be applied to his scientific detective work. He embarked on a process of finding the relevant questions to address.

No theories were available to help Semmelweis find the relevant questions at the time. The same is true now for the challenge of GEO biosafety assessment. For lack of ready-made theory to help us find our way in this technological adventure, we must strive to arrive at an adequate problem definition first. Available biological theories can certainly be of great help in this process. A practical obstacle is that scientific theories are usually not sufficiently applied and context-sensitive to give us clear-cut prescriptions on how to proceed with caution in the applied context of biosafety assessment. Theories by themselves are too far removed from practical applications such as the identification of GEO hazards or the granting of permits for GEO applications. For such purposes, we need specified questions that are relevant to the specific context.

A core quality of scientific research is to raise and isolate questions in such a way that they may lead to reliable answers. Thus, the scientific method is historically an antidote for dogmatic revelation, uncritical intuition and "fragile common sense" (cf. Regal 1990: 87). The crucial activity of asking the right questions has a long history throughout philosophy and science – an important landmark being the wise Socrates, who is arguably among the founders of both.

Logic and general methodology can be seen as conceptual tools for differentiating what is analytically distinct. For example, the methodological imperative of trying to discard confounding factors in conducting or interpreting an experiment, can be seen as the imperative to distinguish between different research questions when we are in danger of mixing them up. If we conduct an experiment in which we cannot *distinguish* the empirical questions that are addressed, then we cannot expect to arrive at an unambiguous answer or empirical claim. For example, if we want to know whether it is *temperature* or *light* that makes a plant grow, then we must separate these questions by varying the relevant factors *independently.* This element of methodological *resolution* in the sense of a distinguishing interpretation of empirical results is a *conditio sine qua non* for scientific research that deserves the name.

All scientific research is centered around raising and answering questions. The quality (in view of our research purpose) of the answers that we hope to get to our scientific questions thus depends on the quality of our process of questioning. Science can be seen as the art of questioning. Research questions can be methodologically inadequate in a given context because they are: *badly chosen in view of the research purpose, inadequately phrased* and/or *improperly addressed experimentally.*

For example, if we study some (biological) process without attempting to exclude possible confounding factors from influencing our observations, the outcome of our research may not really answer our original research question(s). If we fail to recognize such empirical distortions we will end up drawing misguided conclusions from our research. Research assumptions are presupposed answers to implied relevant research questions. Experimental and interpretative rigour and hygiene is a defining element for scientific research and is the essence of the general methodology of science. Safeguarding the quality of scientific methodology is a primary concern of all scrupulous scientists.

Mellon and Rissler (1995) of the *Union of Concerned Scientists* (UCS) have charaterized inadequate biosafety tests as cases of, "don't look, don't find" (Mellon and Rissler 1995: 96). Others have generalized this point by saying that biosafety experiments have so far mainly demonstrated that they were safely planned and *therefore* produced no unwanted effects (Bergmans 1995: 25; cf. NSNE 1997). If so, these experiments are quite useless for the purpose of GEO hazard identification. In terms of this analysis, we can interpret such tests as instances of a *'don't ask, don't answer'* approach. In so far as relevant questions about a possible GEO hazard are not adequately addressed, we will not arrive at reliable answers.

2.3.1 The relevance of research questions

To uncover the scientific kernel of biosafety controversies, the focus in this study is on the research questions that contesting parties choose to put on centre stage in GEO hazard identification. Even in cases where it is difficult to be very specific about the *plausibility of answers* that we find to research questions, it may be possible to be quite decisive about the *relevance of questions* in view of the research purpose. This approach provides a practical tool to analyse biosafety claims and to evaluate the respective relevance of competing claims in a given context of GEO hazard identification. Thus, biosafety controversies are analysed and evaluated as being about *competing sets of relevant questions for GEO hazard identification.* As will be argued in the present and the next section of this chapter, taking general scientific methodology seriously allows us to make rational choices between competing SRQs for a specified research purpose. The presented analysis and evaluation shows that at least some of the long-standing biosafety controversies can be resolved on scientific grounds along these lines.

An analysis of scientific disputes over biosafety assessment in terms of alternative or competing SRQs does not take away the problem, of course, that to some or many questions we may not have scientific answers. This is a difficulty that cannot be overcome scientifically in any other way than by making an effort to recognize those questions and by developing new hypotheses. However, this does not make the present analysis less practical. The comparison of applied SRQs for the purpose of GEO hazard identification still gives us a key to analysing and evaluating contested biosafety claims.

A claim that is not based on the study of relevant research questions is not a relevant claim in relation to a particular research purpose, such as hazard identification of some GEO. In Section 2.4, the problem of deciding about the *relevance* or *irrelevance* of a question in relation to a particular research purpose will be addressed. A claim can be seen as the answer to a question. An empirical claim is the answer to an empirical question. To understand the relevance of a claim in relation to a research problem (such as GEO hazard identification), the relevance of the underlying empirical question in relation to the research problem must be understood first.

Hazard identification of environmental release of transgenic organisms aims to produce a list of relevant questions for the purpose of detecting possible unwanted effects of a GEO. Failure to explicitly address some research question implies that it will not be part of one's list of relevant questions for the purpose (by oversight or by choice). It means that this question will not be considered as (possibly) relevant for the purpose. In other words, it is not included in one's 'window of concern' for that purpose. Apart from addressing or not addressing a question, there should also be concern about the precise *phrasing* or *formulation* of a particular question. Raising a research question in one or another formulation may still lead to different aspects of concern being addressed as more or as less relevant.

Focusing our attention on the applied SRQs at the basis of a particular hazard identification rather than on the available biological theories that pertain to the subject of research, has many advantages for the purpose of the present analysis. One advantage of this SRQ approach is that it allows us to address the peculiarities of a particular research problem very specifically. Theories will typically be phrased in more general terms, leaving us with the problem of interpreting a specific research purpose against the background of a more or less general theory. Thanks to the fact that a SRQ can be precisely adapted to varying contexts and research purposes, it is easier to recognize the special characteristics of a specific problem that is studied. Since hazard identification is typically concerned with recognizing *specific* relevant aspects of a practical research problem, research bias may be expressed in terms of an inadequate SRQ, given the purpose of investigation.

2.3.2 Theories, models, claims, questions

The practical form of applied science is the use of models. Here too, the essence is asking relevant questions. In the practice of science a 'problem definition' will usually get the form of an applied model. However, since there are no ready-made models for the specific purpose of GEO hazard identification either, I choose to concentrate on individual research questions at the basis of a specific biosafety claim. Focusing on models or theories has the *dis*advantage that controversies may take on a paradigmatic character, making a resolution of the dispute more difficult. Since questions are the building blocks of models, it will be easier to evaluate the relevance of individual questions than to evaluate the usefulness of full models. Most parties agree that there is a lack of mature scientific theories or models for the purpose of GEO hazard identification and there is no hope for reaching consensus on this incomplete theoretical basis. Deciding about the relevance of individual research questions is a practical alternative to put real-world problems at centre stage.

According to Haefner (1996), building a qualitative model always begins with specifying an "objective statement", which defines the problem to be understood with the help of the model. Haefner emphasizes that objective statements must be formulated in terms of "goals with purposes". The objective statement must adequately address the problem at hand, otherwise the model will turn out not to be useful for its purpose in the end. Both the goal and the purpose are defining elements of the model pursued, and thus of a sufficient SRQ. If, for example, the purpose of our model would be to *create a genetically engineered organism* rather than to *identify its possible hazards*, then the goal of sufficient understanding would imply different methodological demands.

To do scientific research, one must first specify the research purposes. This is true for empirical research as well as for theoretical research. The best way to do so is to list the questions that need to be answered. Sometimes this is more difficult than it may seem. To design an experiment as well as to build a theoretical model, the questions that are considered relevant for the research purpose must be explicated. To deal with a particular research problem, Haefner (1996) advices, one must begin with an attempt to: "Write down all the questions for which the objective requires an answer. If you cannot do this, then you do not understand the problem" (Haefner 1996: 45). The same issue is addressed by Maynard Smith as follows: "In analysing any complex system, the crucial decision lies in the choice of relevant variables" (cit. in Doucet and Sloep 1993: 306). To recognize the types of questions that must be addressed, inspiration can be drawn from the literature on biological modelling.

A biological model can be seen as a *list of assumptions* with which some problem is interpreted. Constitutive for the assumptions of a model is the *relevance* of the underlying research questions. Thus, individual modelling assumptions can be seen as *hypothetical answers* to assumed *relevant questions*. For example, if one of the assumptions at the basis of a model is that *temperature* will not change, then this implies that the modeller assumes that raising questions about a *change of temperature* could have an influence on the research outcomes. Behind every assumption lies a supposedly relevant question to which the assumption is a presupposed answer. For practical purposes, it is more manageable to discuss and criticize assumptions in terms of the relevance of the underlying questions, given a research purpose.

It is practically (and theoretically) impossible to ask *all* possibly relevant questions. Especially in an applied scientific context such as hazard identification, inclusion or exclusion of possibly relevant questions as part of the applied SRQ must be methodologically justified. As in all scientific research, theoretical models are no less imporant than empirical studies; the former are a prerequisite for a rational design of the latter. In the context of GEO hazard identification specifically, theoretical models may play an additional role since in some cases it may not be *practically* possible to conduct a specific biosafety test (cf. Sukopp and Sukopp 1993: 267). Limits to the feasibility of performing specific biosafety tests may be: *methodological obstacles* such as the need of extremely long periods of research or extremely detailed observations, and *practical obstacles* such as costs, social pressure, ethical considerations, etc.

A scientific claim will always be relative to a more or less explicit SRQ. What qualifies as a 'relevant question' will depend on a given research problem. This implies that a claim about a specific research problem will have to be based on a suitable SRQ given this research problem. Thus, evaluating the relevance of a set of questions as applied in a particular research context provides us with a practical methodological tool to evaluate the relevance of a specific claim in relation to a specific research problem. To evaluate the *usefulness* of a SRQ for a specific research purpose may be less difficult in practice than to assess the scientific *truth* of a claim or a theory.

The analysis of SRQs provides a more practical methodological tool to evaluate contested claims, than an approach of attempting to assess the 'truth' or the 'plausibility' of a specific claim based upon that SRQ or its underlying theory would be (assuming this would be possible in principle). As a method to help us prevent 'overlooking' relevant concerns in hazard identification one cannot rely on existing background theories. Therefore the SRQ underlying a specific claim must be used as an alternative conceptual framework for evaluation. Existing biological theories can 'inspire' in the process of recognizing relevant questions and generating SRQs, but they cannot provide prefab recipes for adequate biosafety tests. Theories can provide inspiration in the process of finding relevant questions of concern for a specific research purpose in the capacity of 'search lights' as characterized by Popper (1968/1934).

Mature theories are typically scarce in the applied scientific context of biosafety assessment. It will often be impossible to produce reliable predictions or claims on expected biological effects, or expected absence thereof. No available biological theory could have predicted something so specific as the effect of genetically engineered *Klebsiella planticola* on the growth of wheat plants. In fact, even though the hazard has now been identified, the underlying biological mechanism is still not completely understood (WWF 1995: 6). Theory allowing predictions about this type of hazard is not available now (and may never be). As Stotzky *et al.* (1993) have argued: "... there are no theories and methodologies available to determine what constitutes an ecologically significant effect on microbial populations and processes in soil or other natural habitats. This lack of appropriate theories and methodologies constitutes a major deficit in microbial ecology and, specifically, in risk assessment of the release of GEMs [genetically engineered microorganisms] to the environment" (Stotzky *et al.* 1993: 95).

By searching for alternative relevant research questions for the purpose of hazard identification, Holmes and Ingham (1994) did manage to recognize the possibility of a hazardous impact. It shows how an analysis of the relevance of research questions for a specific research purpose may be a more practical methodological tool for evaluating contested biosafety claims than relying on the status of biological background theory could be. Problems of scientific bias, for example, can be detected and understood by asking: what would be an appropriate SRQ for the purpose of hazard identification of a particular GEO? Such methodological questions are now mostly left implicit in controversies over possible hazards of applied biotechnology. The evaluation of ongoing biosafety controversies can be improved by focusing our analysis on the explicitly or implicitly applied SRQ for a particular research purpose.

2.3.3 Relevant questions about GEO 'competitiveness'

The following episode from the practice of biosafety assessment may illustrate how opinions about the relevance of questions are formed by existing ideas about biological background theories. Instead of countering one background theory with another background theory, it is more practical to focus on the relevance of individual research questions in view of the specific problem of investigation.

Wrubel, Krimsky and Wetzler (1992) assessed the quality of 30 environmental assessments that were executed by the *Animal and Plant Health Protection Service* (APHIS), which oversees the field testing of transgenic plants as a part of the US *Department of Agriculture* (USDA). Their aim was to, "report what questions are asked (...), what conclusions are drawn, what evidence is cited to support the conclusions, and whether the evidence derives from published or unpublished data, targeted experiments, general principles, or some other source" (Wrubel *et al.* 1992: 281). Although all 30 assessments under study led to a "Finding of No Significant Impact", which gives green light to a proposed release, these conclusions turned out to be not very well supported.

The ecological hazard factor of the *competitiveness* of a released GEO, for example, was not addressed at all in 2 of the 30 assessments and was experimentally evaluated in only 1 of the 30 cases. As for the remaining 27 cases that were analyzed, Wrubel *et al.* (1992) found that questions about competitiveness were mentioned, but not adequately addressed: "No data or discussion are provided or cited for any of the conclusions on competitive ability presented in these 27 environmental assessments" (Wrubel *et al.* 1992: 286). In interviews with the responsible regulators the real cause of this irregular practice was uncovered: "[the] personnel stated they believe that engineered plants carry an added metabolic load (...) that put them at a competitive disadvantage (...). For that reason, they felt that even if seeds from the transformed plants were to escape from the field site, there was little chance the resultant seedlings could compete successfully with wild plants" (Wrubel *et al.* 1992: 286). This conviction led them to omit relevant questions from their considerations and, as a consequence, to make unfounded inferences about competitiveness.

In their recommendations, Wrubel *et al.* (1992) suggest predominantly practical measures to improve the quality of these assessments. It is probably wise to be practical with regard to politically sensitive matters. Practical measures alone, however, will not take away all the obstacles that impair the process of hazard identification. They need to be complemented by more theoretical measures. At issue here is not just the way one specific regulatory institution has been performing. To improve upon the quality of GEO hazard identification we also need to think about what is a methodologically sufficient *set of relevant questions* for the purpose. In my analysis, what Wrubel *et al.* (1992) uncovered was that GEO competitiveness was not adequately included in the SRQ applied by the USDA officials for the purpose of GEO hazard identification. The result was that the USDA review did not adequately address the possible hazard of enhanced competitive ability. The SRQ as applied by the USDA is not suited for its purpose and may thus lead to unrealistic outcomes.

If it is true what Wrubel *et al.* (1992) presume towards the end of their report, that: "[t]he success of genetic engineering for crop improvement depends on public confidence that there is sufficient oversight to minimize the possibility that a transgenic plant might cause adverse environmental effects" (Wrubel *et al.* 1992: 288), then this does not just put demands upon the regulatory apparatus. To win the public confidence will also require establishing a more reassuring image of the science involved. If the general public gets the impression that not all relevant scientific questions are adequately addressed, it is likely to suspend its confidence in the promises of genetic engineering.

The biosafety assessors of the US *Department of Agriculture* were found to work with an unsatisfactory and outdated definition of the possible hazard of GEO 'competitiveness'. Perhaps the most important demand of scientific methodology is to strive for clear and adequate definitions of the concepts and terminology used in research. What definition of 'competitiveness' did the USDA officials have in mind? It seems they interpreted 'competitiveness' as a concept referring to some inherent biological "essence" (cf. Regal 1985) which comes with an isolated entity (an organism). In this interpretation the concept could be meaningfully used in sentences like the following: *'Organism X is competitive'*.

How useful would this interpretation be from a biological point of view? 'Competition' does not make much sense in isolation; other organisms and varying circumstances will also be involved. The notion of competitiveness must be defined as a three-place predicate at least, yielding meaningful sentences like: *'Organism X is competitive with organism Y under circumstances Z'.* This alternative definition of 'competitiveness' addresses a more complex biological situation; it reflects a more inclusive problem definition. For the purpose of hazard identification, this implies that *more questions* become relevant for consideration. Just raising questions about an additional metabolic load of organism X will not suffice to assess its competitive ability, under the more inclusive definition of 'competitiveness'. Besides questions about the transgenic organism that is the focus of a hazard identificiation, questions about the other organisms which a GEO may encounter after release and about the circumstances in which this may happen will also become relevant for the given research purpose. *Alternative definitions of a concept* may produce alternative SRQs for GEO hazard identification (see Figure 2.3).

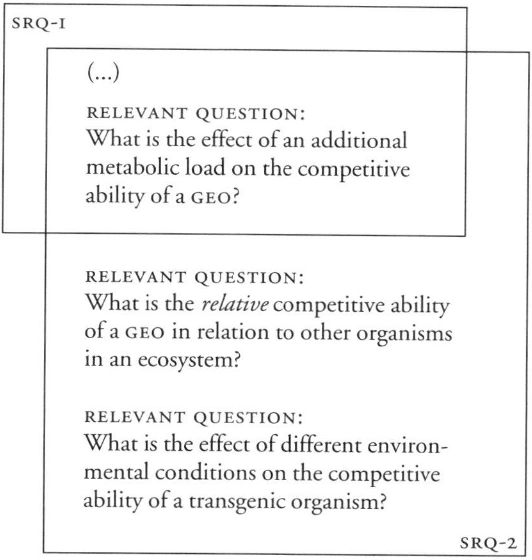

SRQ-1
(...)
RELEVANT QUESTION:
What is the effect of an additional metabolic load on the competitive ability of a GEO?
RELEVANT QUESTION:
What is the *relative* competitive ability of a GEO in relation to other organisms in an ecosystem?
RELEVANT QUESTION:
What is the effect of different environmental conditions on the competitive ability of a transgenic organism?
SRQ-2

FIGURE 2.3: Schematized representation of two alternative SRQs (or 'windows of concern') for the purpose of assessing the hazard of (enhanced) GEO competitiveness. The relevant questions in SRQ-1 have left undetected a possible detrimental effect that can be detected thanks to SRQ-2.

Since adequate SRQs are a *conditio sine qua non* for reliable hazard identification, we must give due care to the adequacy of our definitions. The definition of 'competitiveness' as applied by the USDA was simply not prepared to its task and thus its use gave rise to unfounded inferences. Any *generalizing* claims that a genetically modified organism will be burdened with a so-called "excess baggage" that reduces its competitiveness up to the point of no risk are methodologically flawed, because the implicitly assumed SRQ does not include questions about specific organism-environment interactions as relevant for consideration. Not raising such questions is a sure recipe for overlooking possibly unwanted effects.

— 2.4 The burden of proof in hazard identification —

To study the world we must reduce it to the kernel of what we are trying to understand. The natural scientific tool for complexity reduction is to address specific research questions as relevant and to omit other research questions from consideration. When a SRQ is applied in a practical context, the *burden of proof* will be with the researcher to argue that this SRQ represents the appropriate measure of complexity reduction to be useful for its research purpose. If reasonable doubt is raised about the appropriateness of a SRQ for a specific research purpose, then it will be up to the researcher to justify the claim that his or her SRQ does give a sufficient or useful representation of the world *given the research problem at hand.*

Different research problems will require different SRQs, and different SRQs will be useful for different purposes. This natural burden of proof inherent to scientific research is reflected in the Einstein quote, saying: "Everything must be made as simple as possible, but not one bit simpler" (cit. in Ruef 1997). The scientist who suggests applying a SRQ that leaves out possibly relevant questions in comparison to another SRQ, has the burden of proof for showing that this more simplified SRQ will be equally or more useful for the purpose of GEO hazard identification.

Einstein's phrase can be read as a modern expression of *Occam's razor.* Occam established the methodological imperative that causal explanations should include no more causal factors than is absolutely necessary for our understanding of the phenomenon. This imperative should be observed in the context of applied science also. However, in an applied context such as biosafety assessment there is a serious complication. As yet, we do not know enough about the causal mechanisms active in the subject of investigation. The challenge of biosafety assessment, more specifically of hazard identification, is to arrive at an adequate *problem definition* in view of the research purpose first. Methodological caution does not allow us to exclude possibly relevant questions from a candidate problem definition, unless we can argue that these elements are *not relevant* for the specified research purpose.

This does not imply that *everything* should be included in the considerations until proven superfluous. In practice, the burden of proof for this type of applied science will be on inclusion as well as on exclusion of possibly relevant research questions for identifying GEO hazards. The purpose of the research (identifying hazards and assessing biosafety) implies that in cases of reasonable doubt a possibly relevant question must be *included* in the applied SRQ. Any other methodology of dividing the burden of proof for claims on GEO hazard identification would impair the scientific legitimation of biosafety claims. We want *Occam's razor* to work in applied science too, but we do not want it to carry "a bleeding edge". Application of *Occam's razor* presupposes insight in relevant causal mechanisms; otherwise it comes down to applying a razor in the dark, which is not advisable.

A problem definition in the context of applied science may be seen as a 'window of concern'. It would be *impractical* to expand a window of concern *ad libitum* by including more and more research questions as possibly relevant. On the other hand the metaphor of a *window of concern* clarifies that it would be *imprudent* to make this window narrower than is scientifically warranted.

2.4.1 Reducing complexity

Researchers are faced with a scientific burden of proof for the (ir)relevance of questions. Opposing parties must justify their choice to consider a specific research question as relevant or not. It would be difficult, for example, for anyone to claim now that questions about the role of the *rhizosphere* need not be included in a useful SRQ for the purpose of transgenic *Klebsiella planticola* hazard identification. Those arguing such a thing would thereby take on their shoulders the *burden of proof* for justifying this claim. In any practical research situation, the applied SRQ comes with a burden of proof for arguing its usefulness in relation to the particular research purpose.

The issue of the burden of proof for complexity reduction is not specific to biosafety assessment or hazard identification; it is intrinsic to scientific research generally. Whenever scientists study the world, they cannot but engage in a process of complexity reduction. Any problem definition will always be a reduction of the real world complexity (cf. Luhmann 1986). The choice for one or another form of complexity reduction will require methodological justification. Concerns about unwarranted *scientific reductionism* can be discussed in the practical vein of including or excluding possibly relevant research questions and the associated scientific burden of proof.

The same concerns apply to biotechnology hazard identification. In the process of hazard identification, the complexity of the world cannot but be reduced. This implies a search for the right kind of complexity reduction for the job; *i.e.*, the search for a 'window of concern' which is narrow enough to be practical and wide enough to be prudent. If, for example, the chosen approach excludes certain aspects of biological complexity that are relevant for GEO hazards, then it may create a situation in which possibly important information is overlooked. A misguided type of complexity reduction may lead the way to a misguided type of conclusions. For practical reasons it may be preferable to work with the most reduced 'window of concern' – but not at all prices.

This implies, for example, that those who would argue that 'competitiveness' can be meaningfully interpreted and applied as a one- or two-place predicate, thereby take on the methodological burden of proof for excluding other considerations ('predicates') as not relevant for the purpose (cf. Section 2.3.3). Failure to address questions about the environment and about competing organisms as relevant concerns, would lead to a *methodologically flawed* basis for biosafety claims on GEO competitiveness. Along the same methodological lines, all problem definitions and SRQs can be evaluated as more or less sufficient for their purpose.

Decisions about whether a specific research question should or should not be included in a SRQ for a given purpose, are essentially comparisons between two possible SRQs. Before a question can become part of an accepted SRQ, those who wish to include that specific concern must provide reasons why they think the suggested question(s) is (are) relevant and should be part of the SRQ for the given purpose of GEO hazard identification. If it does not become clear in what way addressing a specific research question could contribute to an improved understanding of the problem at hand, then the suggestion must be rejected and the contested concern need not be made part of the SRQ. Thus, *including* as well as *excluding* research questions as relevant part of a SRQ comes with a methodological burden of proof. In accordance with *Occam's razor*, the notion of a SRQ implies the intention to work with a minimal but sufficient set of relevant questions. Thus, individual research questions can be weighed as being relevant for inclusion in an appropriate SRQ for a specified research purpose, or not (see Figure 2.4).

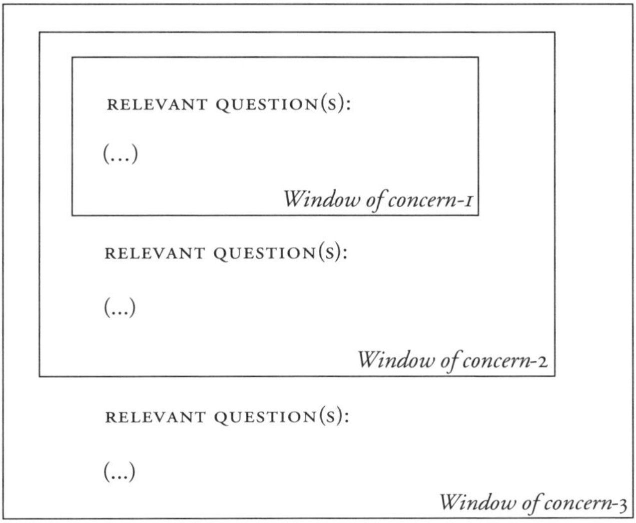

FIGURE 2.4: Schematized representation of the scientific burden of proof for inclusion or exclusion of (a) specific research question(s) as relevant part of an SRQ. Both expansion and reduction of a window of concern must be argued for. This does not imply that a 'wider' window of concern will always be the better one. For practical purposes, a window of concern should be kept as limited or narrow as possible, without loss of relevant detail. This is in accordance with *Occam's razor* and with Einstein's imperative: "Everything must be made as simple as possible, but not one bit simpler" (cit. in Ruef 1997).

Considering the importance of applying a *wide* and *narrow* enough problem definition for the purpose of biosafety assessment, I suggest to apply the following practical rule for dividing the burden of proof for using a more or less inclusive SRQ: *as long as the argument for excluding a question is not decisive (given an argument to include it), the specified contested question of concern should be considered relevant for the research purpose.* The rationale for this approach is that an unwarranted exclusion of a possibly relevant research question may have serious unwanted consequences. This *scientific* burden of proof for the methodological basis of a biosafety claim does not take away the possibility for a society to decide on *political* grounds to use a less inclusive SRQ as a sufficient scientific basis for biosafety assessment (the asymmetry between a *scientific* and a *political* burden of proof for legitimating biosafety claims will be discussed in more detail in Chapter 3 and in Chapter 6).

Taking the burden of proof in the process of hazard identification seriously, it may be used as a tool to choose between competing SRQs for a specific research purpose. If one assumed SRQ includes a possibly relevant concern for hazard identification which is not included in an alternative assumed SRQ, then the latter SRQ implies a greater risk of overlooking possible hazards than the former (all other elements of the studied system being equal). In such a case, *the purpose of hazard identification* (not overlooking possible hazards) requires one to choose the more inclusive SRQ as a more appropriate basis for investigation. This situation may change as soon as the contesting party produces enough 'proof' to argue that the contested concern *need not be included in a SRQ*. On that new basis the method of hazard identification may be changed in agreement with the less inclusive SRQ.

Different purposes will require different SRQs. A biotechnology engineer attempting to insert a gene and have it expressed in a transgenic organism will apply a different SRQ than a scientist will need whose objective it is to assess the safety of a GEO. In this respect I agree with Van den Daele (1992) who has expressed this asymmetry as follows: "The engineer who is asked to implement a technology may have an easier task than the ecologist who is asked to assess the environmental risks of that implementation. The engineer is looking for appropriate causes. He must answer the question, 'what can we do to make X happen?' Such questions are restricted in scope, and they correspond to the standard explanatory strategies in experimental science which try to establish cause-effect (if-then) relationships. (...) The ecologist by contrast is looking for consequences and functions. He must answer the question, 'what will happen if we do X?' Such questions are unrestricted in scope, highly complex and less well elaborated in the existing body of knowledge" (van den Daele 1992: 331).

If a technological option works without further additions (without asking more questions), then this may be a sufficient basis for engineering purposes. In the context of *hazard identification* and *biosafety assessment* (as well as in technology assessment more generally), however, there is not such a clear-cut test as *'does it work?'* for evaluating the relevance of research questions and the sufficiency of an SRQ. Thus, an asymmety in the burden of proof for more or less inclusive SRQs exists between *biotechnology engineers* on the one hand and *biosafety assessors* on the other hand.

2.4.2 Experimental design produces empirical evidence

The relevance of research questions implies that *experimental data* and *scientific evidence* will be relevant *or not* in a certain context, depending on whether they are the empirical result of applying a suitable SRQ for the research purpose (cf. Krimsky 1996; Krimsky *et al.* 1995). In terms of the present analysis we could say that a "microcosm study" is the experimental materialization of a SRQ. The applied SRQ specifies the addressed research questions and thus it specifies for which specific biosafety claim the experimental results acquired from the study of some specific *microcosm* can count as 'evidence' for that claim. To give an empirical basis to a biosafety claim, the research questions that are methodologically relevant for that claim must be recognized and addressed.

For a succinct interpretation of some of the scientific terms involved, consider the definitions given by Haefner (1996) in *Modeling Biological Systems.* Especially important for my analysis of scientific complexity reduction is his notion of a "well-defined system", which is defined by Haefner as: "The smallest set of objects and relations whose states (values) cannot be proved to be unnecessary to achieve the objectives of the model" (Haefner 1996: 13). In his interpretation, "a model" (noun) is, "a description of a system" and, "to model" (verb) is, "the human activity of creating a description of a system". A "system" in Haefner's definition is, "a collection of objects and relations between objects". What Haefner calls a "well-defined system" is an abstract representation of the biological complex under study; it is the result of scientific complexity reduction.

This definition of a "well-defined system" is especially useful for the present analysis because it includes the important issue of the burden of proof that is involved with claiming to have a good enough picture of a biological system. We must provide arguments for leaving out system ingredients as unnecessary. This is the general burden of proof that is involved with the process of scientific complexity reduction. Of course, the notion of a "well-defined system" also implies that the constituting theoretical elements of the model are clearly defined. Carefully observing the latter general methodological prerequisite can help us to avoid much of the fruitless interpretative confusion that is too often at the basis of biosafety controversies (cf. van Dommelen 1995).

Since the possibility for *objects* and *relations* to develop over time is important for the purpose of prospective biosafety assessment, the consideration of *change* must also be included as an essential component of a "well-defined system". Taken together, there are three necessary types of components to construct a "well-defined system" and a "useful model": 'the relevant system objects', 'the relevant system relations', and 'the relevant system changes' (cf. van Dommelen 1998). It should be noted that one can never be completely certain to have indeed included a sufficient set of system components or relevant questions for detecting or identifying all possible hazards.

An important aspect of understanding the burden of proof involved in scientific research is to understand the methodological basis of experimental evidence. The example of the alternative biosafety tests for transgenic *Klebsiella planticola* illustrates how the methodological status of empirical results is totally dependent on the way the 'evidence' was collected. Knowingly or unknowingly omitting relevant system objects, and/or relevant system relations, and/or relevant system changes from concern, will lead to biased and methodologically flawed scientific evidence.

This implies that empirical data carry no information unless they are interpreted against the background of the specific design of the experiment that produced them. Using sterile soils for hazard identification of transgenic *Klebsiella planticola* produced 'experimental evidence' that genetically engineered *Klebsiella planticola* poses no threat to the growth of wheat plants. An alternative experimental design (including the rhizosphere as an element of consideration) produced completely contradictory empirical results. The only way to evaluate the contradictory empirical findings is to evaluate the two respective SRQs that were used to generate these results in view of their research purpose. Insufficient attention for the methodological role of the assumed SRQ in the design and in the interpretation of experiments, may give rise to practically void empirical claims.

Since experiments are intended to answer our research questions, adequate care for the design and interpretation of these experiments is a prerequisite for drawing meaningful conclusions from empirical research for biosafety assessment. An assumed SRQ which is not useful for its research purpose will not yield relevant data. Regal (1994) has characterized such would-be empirical findings in relation to GEO hazard identification as "nondata on nonreleases" (Regal 1994: 11). This is a major concern in biosafety controversies, because the antagonists in a debate will typically refer to "empirical data" and "scientific evidence" to support their claims, often without any methodological clarification of their relevance. This tendency adds to the rise and persistence of *artificial* as opposed to *fundamental* controversy.

2.4.3 Relevant questions about ecological 'time scales'

Consider the following example to illustrate the relevance of experimental design for the evaluation of empirical claims. Crawley *et al.* (1993) have described a field experiment for assessing the *invasiveness* of transgenic plants. 'Invasiveness' was defined in this study as the rate of increase of different genetic lines under a variety of experimental conditions. The research object was, "to determine whether genetic engineering for herbicide tolerance affects the likelihood of oilseed rape becoming invasive of natural habitats" (Crawley *et al.* 1993: 620). Values of the "invasion criterion" larger than 1 predict that the plant will increase in abundance under the given set of environmental conditions; values smaller than 1 predict that the plants will decline to extinction. On the basis of their research over a period of three years, from 1990-1992, Crawley *et al.* (1993) concluded that, "there is no evidence that rape is invasive of undisturbed natural habitats, and no evidence that transgenic lines of rape are more invasive of, or more persistent in, disturbed habitats than their conventional counterparts" (Crawley *et al.* 1993: 622).

Reassuring as experimental evidence of this kind may seem, the most important question to ask is: how must these data be interpreted? Since biosafety claims can only be evaluated against the background of the SRQ upon which they are based, a closer look at the assumed SRQ that Crawley and colleagues used as a basis for their experimental design is required. In a detailed analysis, Weber (1995) has called attention to "nicht beantwortete Risikofragen" in the research by Crawley *et al.* (Weber 1995: 118). The methodological status of experimental data can only be assessed on the basis of the research questions that were addressed.

Kareiva (1993) has called the publication in *Nature* of the findings by Crawley *et al.* (1993) a "landmark paper" because it is an attempt to produce quantitative experimental results, which are sorely missing in the biosafety debate and because their experiment is, "one of the most comprehensive population studies ever undertaken in plant ecology" (Kareiva 1993: 580). However, Kareiva's praise is seasoned with irony and he proceeds to downplay the relevance of these empirical results: "too much should not be read into these results, however" (Kareiva 1993: 581). The general basis for his skepticism about the relevance of these data is in the fact that, "history tells us that an ultimately succesful invader might initially fail miserably, or barely persist for decades" (Kareiva 1993: 581).

Interpreted in terms of an adequate SRQ for identification of the hazard of invasiveness, we can say Kareiva (1993) argued that Crawley *et al.* (1993) have not addressed sufficient questions concerning a long enough time-scale in their experiments to generate fully relevant results. If *invasiveness* does not occur after two or three years (which was the time scale included in the study by Crawley *et al.*), then this does not tell us much about whether it will or will not occur ten or fifty or hundred years later. Kareiva shows us that Crawley's claim is not very relevant for lack of an appropriate SRQ as a methodological basis of his experiment (see Figure 2.5).

The assumed SRQ applied by Crawley and colleagues in this experiment for the identification of GEO invasiveness does not include a proper *time scale* for the research purpose and thus cannot give scientific legitimation to claims about the possibility of enhanced invasiveness over a longer period of time. The possibility of invasiveness developing in a later phase is overlooked by the experimental design based on a SRQ which excludes this type of ecological change over a longer term. The alternative assumed SRQ implied by Kareiva's critique does take into account the relevance of considering longer periods of time for the identification of the hazard of invasiveness. Summing up, Kareiva notes: "This is because year-to-year variation in plant success can be staggeringly large – a phenomenon well-known to any gardener" (Kareiva *et al.* 1996). The example shows that concerns about 'system change' should not be overlooked in experimental studies of 'system objects' and 'system relations'.

Since Crawley's assumed SRQ (in this experiment) is more reduced in this respect than Kareiva's assumed SRQ (and given Kareiva's arguments for applying a more inclusive SRQ), the scientific burden of proof would be with Crawley and colleagues to argue that longer periods of time *need not be considered for the purpose of GEO invasiveness identification.* If ecological impact may remain 'hidden' over a shorter period of time, then a biosafety claim on the absence of effects in the longer run cannot be given a sufficient methodological basis if it is restricted to shorter term field-testing. From a later publication (Crawley 1994) it seems Crawley will not be inclined to take on the burden of such a proof. There he does more justice to the relevance of considering longer periods of time for the purpose of identifiying invasiveness and cites an example of a population not breaking the "nothing is happening threshold" until 100 years after the release (Crawley 1994: 32).

QUESTION:

What is the effect of longer ecological time scales on the invasiveness of transgenic organisms?

SRQ-1	SRQ-2
Arguments for *excluding* this question from SRQ as NOT RELEVANT for assessment of GEO *invasiveness*:	Arguments for *including* this question in SRQ as RELEVANT for assessment of GEO *invasiveness*:
"Lack of data to interpret long-term dynamics"	"Year-to-year-variation makes interpretation of short-term experiments difficult"
"Practicality of short-term experiments as opposed to long-term experiments"	"Ecosystem changes may develop over longer time scales, making short-term studies of invasiveness less relevant"

FIGURE 2.5: Schematized representation of the scientific burden of proof for inclusion or exclusion of a specific research question as relevant part of an SRQ for the hazard identification of GEO invasiveness.

— 2.5 Evaluating contested claims on 'pathogenicity' —

As a demonstration of the analytical and evaluative approach developed in this chapter, consider the following opposing claims on the hazard of *inadvertently converting a non-pathogen into a pathogen* by genetic engineering. Identifying the hazardous potential of GEO pathogenicity has given rise to one of the long-standing disputes in GEO biosafety assessment. My analysis and evaluation of this controversy is meant to serve as a demonstration of the possibility to resolve seemingly "endless technical debates" (cf. Section 1.4.1) *on scientific grounds* by interpreting the dispute in terms of competing sets of relevant questions and the associated scientific burden of proof for complexity reduction.

In 1987, the US *National Academy of Sciences* (NAS) issued a report of its *Committee on the Introduction of Genetically Engineered Organisms into the Environment* (NAS 1987). One of the arguments in the report concerns the possibility that engineered organisms might inadvertently cause disease:

> "Among the dangers envisioned in r-DNA genetic engineering of microorganisms is the inadvertent conversion of a nonpathogen into a new, virulent pathogen. How valid is this concept? It is important to recognize that virulent pathogens of humans, animals, and plants possess a large number of varied characteristics that in total constitute their pathogenic potential. The traits contributing to pathogenicity include the ability to attach to specific host cells, to resist a wide range of host defense systems, to form toxic chemicals that kill cells, to produce enzymes that degrade cell components, to disseminate readily and invade new hosts, and to survive under adverse environmental conditions outside the host. Together with the need to compete effectively with many other microorganisms for survival, these traits form an impressive array of requirements for pathogenicity. *The possibility that minor genetic modifications with r-DNA techniques will inadvertently convert a nonpathogen to a pathogen is therefore quite remote ...*" (NAS 1987: 15 – italics added).

The NAS report suggests that, given the "impressive array of requirements for pathogenicity", genetic engineering of nonpathogens in effect poses no hazard of creating a pathogen. This implies that in a hazard identification of a genetically engineered nonpathogen, no questions need to be asked about possible pathogenicity.

The practical consequences of this particular biosafety claim for biotechnology regulation can be far-reaching. Davis (1989), for example, has concluded along this line of reasoning: "Instead of insisting, as a number of ecologists have done, that we require extensive tests of every genetic novelty, on a case by case basis, for possible spread and harm, we know enough to justify exempting recombinant products of nonpathogenic bacteria" (Davis 1989: 865; cf. Brill 1985, Beringer and Bale 1988). This means that, according to Davis and others, in the process of GEO hazard identification there is no need to ask questions about the possibility of converting a nonpathogenic bacterium into a pathogen, because this is a remote possibility. The hazard of pathogenicity is interpreted here in such a way that its complex character makes it unnecessary to be concerned. What remains as a relevant concern is the question whether the original host organism is considered pathogenic or not. This means that the NAS and others think that the identification of the hazard of pathogenicity does not need experimental testing at all, as long as the original organism is a nonpathogen.

Not everyone in the debate agrees with this assessment. Sharples (1991), for example, has contested this biosafety claim put forward by NAS and others. In her view, "... small genetic changes do not invariably translate into small changes in ecologically relevant aspects of phenotype. There is not necessarily a proportionality between the two (...). From the ecological point of view, the number of genes, added or subtracted, is not an appropriate standard by which to judge the necessity for risk analysis" (Sharples 1991: 21). Elsewhere she argued: "Many instances of drastic change in the ecological, medical, and economic status of diverse kinds of organisms are known to have been caused by changes in single genes. Examples include antibiotic resistance in bacteria, pesticide resistance in insects, and changes in the pathogenicity of viruses and fungi. This is not to say that every single gene has this potential. Still, it is not the number of new genes that can be of ecological significance, but rather which genes and what their functions are that is important" (Sharples 1987: 95; cf. Pimentel *et al.* 1989).

Biosafety assessors must indeed cope with a lack of scientific knowledge. However, there is enough information available now to decide that some questions cannot be left out of consideration as being *not relevant* for the particular research purpose of identifying the hazard of *pathogenicity*. As demonstrated below, it is possible to take a scientifically based stance in this controversy.

2.5.1 Defining 'pathogenicity'

Thinking in terms of a sufficient SRQ in view of the research purpose, raises doubts for several reasons. In the first place, hazards are not limited to disease, and causes of infectious disease are not limited to pathogens. The latter point is obvious from the following, common definition of "pathogen":

> "[A pathogen is] any microorganism which, by direct interaction with (infection of) another organism (by convention, a multicellular organism), causes disease in that organism. A pathogen is not, therefore, any microorganism which is causally connected with disease: for example, a microbe which produces a toxin that causes disease in the absence of the microbe itself would not be regarded a pathogen".

A case in point is insect control through pest-resistant plants. Plants can be made resistant to insects, for example, by taking the genes that produce delta-endotoxin in *Bacillus thuringiensis*, inserting them in strains of *Pseudomonas fluorescens* and releasing the latter in the rhizosphere of plants (Atlas 1985: 223). According to the definition, *Bacillus thuringiensis* is not a pathogen; it is already widely used for biological control of pest-insects and is considered safe for many agricultural applications. Delta-endotoxin can cause disease in insects, but since it is not an organism it cannot be considered pathogenic. Also, the engineered strains of *Pseudomonas fluorescens* can cause disease in insects, but, since they can only do so indirectly (through the activity of delta-endotoxin) they are not considered to be pathogenic. Although the original strains of *Pseudomonas fluorescens* cannot cause disease in insects and the engineered strains can, this cannot be considered as the conversion of a nonpathogen into a pathogen. Also, should the ability to produce delta-endotoxin inadvertently be conferred to another micro-organism, which thereby becomes capable of causing disease in insects, this would not be considered as an inadvertent conversion of a nonpathogen into a pathogen.

The point of hazard identifications that focus on possible pathogenicity in engineered organisms is that we do not want the organisms to cause undesirable diseases. The emphasis in the passage quoted from the NAS report on pathogenicity narrowly defined is therefore misleading. Nonpathogens can potentially cause disease in many different ways and this should be reflected in the applied SRQ for GEO hazard identification.

Generic safety arguments envisaging nonpathogenic organisms are also problematic because it is hard to draw the line between pathogens and nonpathogens. A full-fledged definition of 'pathogenicity' would show that we are dealing with a context-dependent feature. Organisms are not pathogens or nonpathogens simpliciter. *An organism X is a pathogen with respect to organism Y in conditions C if X has certain effects Z on Y in C.* 'Pathogenicity' is therefore a *four-place* predicate covering a complex relationship, not a simple one-place predicate representing a non-relational feature. Hence organisms known to be nonpathogens in specific circumstances that have been subject of investigation may well turn out to be pathogens in different circumstances. A more-place predicate (such as pathogenicity) implies that *more* questions are relevant in its 'identification' or assessment and thus that a more inclusive SRQ should be used for hazard identification.

For example, the pathogenicity of a micro-organism can depend on the condition of a host. A so-called 'opportunist pathogen' will not be pathogenic to a specific host when the latter is in a 'normal' condition. The opportunist may become pathogenic, however, as soon as the host becomes weakened for some reason. This is a serious complication in the assessment of pathogenicity. Since an 'opportunist pathogen' is not pathogenic under normal circumstances and becomes pathogenic when the host is weakened, we can say in such a case that a nonpathogen changes into a pathogen. Since it is conceivable that one engineered trait in an opportunist pathogen might give it the ability to weaken its host (by producing a toxin, for example) and thereby to become pathogenic to that host, the statement that "minor genetic modifications" will not convert a nonpathogen into a pathogen is misleading. "Minor genetic modifications" may also lead to pathogenicity through increased virulence, and they may cause a broadened host range (Adelberg 1985), such that a pathogen may come to cause disease in new species.

Remember that the Committee remarks that pathogenicity results from an "impressive array" of required traits: the ability (1) to attach to specific host cells, (2) to resist a wide range of host defense systems, (3) to form toxic chemicals that kill cells, (4) to produce enzymes that degrade cell components, (5) to disseminate readily and invade new hosts, and (6) to survive under adverse environmental conditions outside the host, together with (7) the need to compete effectively with many other microorganisms for survival.

The NAS (1987) argues that minor genetic modifications will not transform a nonpathogen into a pathogen because pathogenicity involves so many different traits. As it stands, the argument is unsatisfactory for the following reasons.

First, if a micro-organism is artificially given one of the traits, say the ability to attach to specific host cells, this may well be the starting point for the organism to acquire the other traits through natural selection.

Second, the argument assumes that nonpathogens do not possess *any* of these traits. This assumption is entirely unfounded. Why could not some nonpathogens miss only one trait, a trait which is easily introduced in artificial modification, by design or by oversight? If the only trait missing is the ability to "resist a wide range of host defense systems", then a mutation introducing this ability would convert a nonpathogen into a pathogen. Indeed examples of this are known. Strains of *Agrobacterium tumefaciens* differ in their host ranges due to slight genetic differences. The wide host range (WHR) strain cannot infect some plants that the limited host range (LHR) strain can infect; WHR bacteria provoke a hypersensitive response which confers resistance on the plant (Nester *et al.* 1985). As Colwell (1985: 231) notes, a mutation deleting the ability to provoke the response would make the WHR strain pathogenic to previously resistant plants.

Third, the argument assumes that the changes in traits needed for pathogenicity to arise require *genetic* changes. This assumption is obviously false. For example, changes in the environment may well affect competitive ability in an organism (cf. trait 7).

QUESTION: What impact of a GEO may confer pathogenic effects?	
Relevant because... "GEOs may have new pathogenic impacts on other organisms"	*Not relevant because...* "Stick to what is known about pathogenic agents now"
QUESTION: Do different environmental conditions affect GEO pathogenicity?	
Relevant because... "Absence of pathogenicity may be artefact of environment"	*Not relevant because...* "Impossible to study all environmental variations"
QUESTION: Which organisms could be affected by a pathogenic GEO?	
Relevant because... "Impact on seemingly unimportant organisms can backfire"	*Not relevant because...* "Restrict concern to organisms of economic interest (*e.g.*, crops)"
QUESTION: (...)	

FIGURE 2.6: Schematized representation of the scientific burden of proof for inclusion or exclusion of individual research questions as relevant part of an SRQ for the hazard identification of GEO 'pathogenicity' (no claim for *completeness* implied).

These observations may suffice to demonstrate that the concept of 'pathogenicity' and its biological reality are so complex that we need quite an extensive SRQ to identify its possible occurrence in the application of a GEO. Those claiming that fewer questions will suffice, thereby take on the burden of proof for this complexity reduction (see Figure 2.6). With the cited biosafety claim, the NAS takes on the scientific burden of proof for the exclusion of the questions mentioned above from a SRQ for the purpose of identifying the hazard of GEO pathogenicity.

Given the preceding considerations, it becomes clear that the claimed impossibility of an unwanted or unexpected conversion of a nonpathogen into a pathogen, is not an *outcome* of any research but rather an *input* of it in the form of the choice of a restricted SRQ. If the NAS or others fail to address the relevant questions for this research purpose, they cannot expect their claims to be relevant to the specified research problem.

— 2.6 Developing sets of relevant questions —

As was shown in the previous sections, what is considered an appropriate SRQ for a particular research purpose may evolve. What was first considered a sufficient SRQ for identifying the hazard of GEO invasiveness by Crawley *et al.* (1993), for example, was criticized by Kareiva (1993), and was later adapted in accordance with Kareiva's critique. Thus, an applied SRQ for the purpose of GEO hazard identification may be expanded or reduced in response to scientific criticisms and developing insights.

The scientific objective of biosafety assessment should be to avoid methodological pitfalls and to recognize that, "... the absence of evidence for environmental risks may well be more indicative of 'absence of looking' rather than absence of risks" (Kareiva and Parker 1994: 8). Just as future scientific results may change the division of the burden of proof in relation to the contested relevance of a specific research question, the ideas about a suitable SRQ for the purpose of a specific GEO hazard identification may also change.

This developing nature of SRQs is in accordance with the function of the 'flowcharts' (see Section 2.1) that I consider to be examplary attempts at specifying the questions of a biosafety test. The ABRAC (1995a, 1995b) intends the flowcharts, "to accommodate a dynamic, rapidly changing state-of-the-art" (ABRAC 1995a: 2). This recognition of hazard identification as a *learning process* connects well with an important concept from the context of international biosafety policy. At the platform of the *United Nations Environmental Programme* (UNEP), there is concern about the necessity of international *capacity building* in relation to biotechnology risk assessment (cf. UNEP 1995: 16). In my view, this concept applies to all countries of the world. Both the Southern *and* the Northern countries face the challenge of developing our knowledge for GEO biosafety assessment. In fact it could be argued that, since a large part of the world's *biodiversity* is found in Southern countries, these countries should have a large scientific and political say in the design of appropriate procedures for biosafety assessment.

It seems to me that it would be a wonderful asset if such practical and adaptive tools for biosafety testing as the ABRAC 'flowcharts' (which are already available as computer software) would be made available on the world-wide *Internet* and become the basis of a global network of biosafety expertise. Preliminary steps in this direction are already taken by 'internet conferences' on the biosafety of genetic engineering and developing *web sites* on the subject (cf. *Dutch Ministry of the Environment, European Union, US Environmental Protection Agency*, etc.). Such an integrating and up-to-date global resource of relevant research questions and empirical results would be useful for all *Competent Authorities* around the world to find support in their difficult task.

A *web site* of this kind could become a lively arena for discussing the inclusion or exclusion of research questions for biosafety testing and for developing useful sets of relevant questions for the purpose of GEO biosafety assessment. The biological hazards of GEOs are to a large extent a global affair which does not stop at nation boundaries (cf. Rissler and Mellon 1993; cf. Regal 1997). Therefore, the identification of these hazards should be made a global effort and concern as well (for more detail, see Chapter 5).

Controversies are potentially very fruitful *resources of relevant questions* in this process of scientific development. Although the burden of proof for the sufficiency of a specific SRQ lies with scientific researchers, suggestions for candidate relevant questions can be made by the general public also. This means that the *public perception* of biotechnology hazards can also be given a place in the process of scientific biosafety assessment as one of the sources from which awareness about possibly relevant questions may arise, thereby giving substance to the input of *public participation*. An important advantage of a focus on the relevance of research questions is that *raising questions* is not a methodological privilege of scientists alone. Relevant questions may be and are being raised by lay audiences as well. On this basis of a developing international network of 'hazard detectives' it remains up to the scientists involved to decide about the inclusion or the exclusion of candidate relevant question as a part of the applied SRQ.

The analysis presented in this chapter does not take away the fact that politicians may decide *on political grounds* to overrule the scientific burden of proof and to go ahead (or not) with a specific GEO release *despite the outcome of a scientific biosafety assessment.* In practice, politicians have the power to decide that social interests legitimate acting without a sound scientific basis. For example, if it were assumed that the use of modern biotechnology is an inescapable means for feeding the growing world population, then this could be a legitimate political reason for accepting possible environmental risks. This would mean that the *science* of an assessment is overruled by the *politics* of an assessment (for more detail, see Chapter 6). An important concern is that such a political decision should not be presented or interpreted as a *scientifically based* green light. The latter can only be given in so far as the scientific burden of proof for complexity reduction is respected.

CHAPTER 3

Hazard Identification of Herbicide Resistant Plants

— 3.1 A political experiment —

In Chapter 1, a policy-maker was introduced who found himself in an awkward position: having to develop regulation on a subject about which the scientists are in disagreement. Given the possibilities of genetic engineering, we would like to foresee inadvertent consequences. According to Krimsky (1996), "[t]he creation of crops that are resistant to herbicides is among the most controversial applications of biotechnology to agriculture" (Krimsky 1996: 29). To what extent will the technology be safe for humans and the environment?

In Germany this controversial issue of transgenic herbicide resistance in agricultural crop plants was put at centre stage of an extensive *technology assessment* (TA) funded by the *Ministry for Research and Technology*. From February 1991 to June 1993, a group of about 60 representatives from industry, environmental groups, regulatory agencies and research institutes joined forces in a TA of genetically engineered herbicide resistant crops. Under the coordination of the *Wissenschaftszentrum Berlin für Sozialforschung*, they produced and discussed eighteen 'expert reports' about possible costs and benefits of genetically engineered herbicide resistant plants and the associated use of new herbicides in agriculture (cf. van den Daele *et al.* 1996: 323-324).

In an evaluation of the TA process, project coordinator Van den Daele calls this TA study a "political experiment" in which, "[t]he bringing together of experts with different political convictions guaranteed that the unavoidable 'softness' of expert opinions and scientific controversies, where they existed, could be brought to light" (van den Daele 1994: 6). In my view, this is an overly optimistic understanding and it disregards serious limitations of this TA process.

Bringing to light the "softness" of expert opinions and scientific controversies is indeed a prerequisite for democratic decision-making and for avoiding that political decisions are presented in a scientific guise. Van den Daele's optimistic assessment draws on his general understanding of the potential of TA studies: "The very idea behind the TA is that it involves the possibility that handy claims, by means of which one can beat the drum for a technology in public or provoke resistance to the same technology, can be shown to be unprovable, poorly justified or simply wrong" (van den Daele 1994: 5). The aim of this chapter is to uncover how exactly those "handy claims" were shown to be "unprovable, poorly justified or simply wrong" in this TA study and to present the outlines of a more productive alternative. Along the lines of the analysis presented in Chapter 2, the way scientific expertise was incorporated in this case of technology assessment will be critically assessed.

The project coordinator of the German TA study emphasizes the importance of scientific knowledge and empirical information as an input in the biosafety assessment of herbicide resistance technology (HR technology): "The findings themselves, were supposed to be knowledge not politics. They were a matter of insight, not of interest" (van den Daele 1994: 7; van den Daele *et al.* 1996: 10). The accepted common ground of the herbicide resistance technology assessment (HR-TA) was that the answers to its questions, "did not depend on political or moral evaluations, but on science" (van den Daele 1994: 117). Thus, the main substance of debate in the course of the TA process were the "empirical controversies" (van den Daele 1994: 117) and in the final meeting the discussion centered around "argument trees" and "empirical conclusions" in which the managers of the TA study had, "summarized the results of the scientific discussions in the TA study" (van den Daele *et al.* in Böger 1994: 127). As a general program of technology assessment, it would be difficult to find fault with this approach. My concern here is not so much with the program, but rather with the scientific methodology underlying the assessment process.

In relation to the method of, "summarizing the results of the scientific discussions", the following statement[1] is made: "Die Frage ist natürlich, welche Schlußfolgerungen man vorgibt, um Einigung oder Widerspruch zu erzeugen. Die Antwort kann nur sein: die nach der Sachlage plausiblen Schlußfolgerungen"[2] (van den Daele *et al.* 1996: 28). In my view, this statement does not so much give a *solution to the problem* but rather only *states the problem.* We are left with the same question that we encountered as a hurdle to take at the end of Chapter 1: what do we consider to be the 'plausibility' of a contested claim in this context? To answer this question, we cannot get around producing a view on the *role of science in biosafety assessment* and on the *interpretation of controversy*

1 Translations in footnotes are by AVD unless otherwise indicated.

2 "The question is, of course, which conclusions one uses to arrive at consensus or dissensus. The answer can only be: the conclusions which are most plausible in the given context".

in applied science. The HR-TA that is discussed in the present chapter does not give a clear view on either of the two concerns.

In his introduction to the final report, the TA coordinator writes: "Ob das Experiment gelungen ist, müssen unbefangene Beobachter beurteilen. Der hier vorgelegte Endbericht wird die Öffentlichkeit in die Lage versetzen, sich selbst ein Bild von den Ergebnissen, den Leistungen und den möglichen Schwächen des TA-Verfahrens zu machen"[3] (van den Daele *et al.* 1996: 32).

In the end, the German HR-TA study did not live up to the expectations raised. The participative design of the study was frustrated when part of the involved experts decided to walk out of the final meeting. The main reason for its limited success, as will be argued below, has been the lack of a clear view on the role of science in biosafety assessment and on controversies in applied science. On the other hand, the HR-TA was a success in so far as it produced a wide array of expert opinions in the form of eighteen detailed reports. This material remains available for re-evaluation.

This TA process would have had a better chance of success, if 'artificial controversies' had been explicitly recognized as an obstacle for scientific discussion. A focus on the *relevance of research questions* given the purpose of technology assessment, could have stimulated fundamental discussions and a constructive outcome.

— 3.2 Experts withdraw from deliberations —

The head of the coordinating institute and the main project coordinator of this TA procedure, Van den Daele, writes about its general approach: "Das *Wissenschaftszentrum* hatte den Auftrag, sämtliche Informationen und Argumente so zusammenzufassen, daß auf der Abschlußkonferenz daraus Schlußfolgerungen gezogen werden können. Zu diesem Zweck wurden zunächst die Aussagen aller Teilnehmer zu einzelnen Themen zusammengestellt. In einem zweiten Schritt wurden die Argumente, Gegenargumente, Belege, Einwände, etc. zu zentralen Zielbehauptungen logisch geordnet. Die Antragsteller des Verfahrens haben vorläufige Schlußfolgerungen gezogen"[4] (van den Daele 1994 – appendix: 40;

---

3 "Whether the experiment has been a success, must be decided by independent observers. The present final report will enable public discussion to develop a picture of the results, the achievements and the possible weaknesses of the TA process".

4 "The *Wissenschaftszentrum* had the task to summarize information and arguments in such a way that final conclusions could be drawn at the final meeting. For this purpose, the claims of all participants were first ordered in individual themes. In a second step,

van den Daele 1996). This general approach fails to contribute to the resolution of controversy, and it falls short of showing why claims are, "unprovable, poorly justified or simply wrong". Summaries and orderings of debates are necessary in a TA process, but we need more than that. The primary objective of technology assessment is to identify the *relevant questions* of concern; this has not been done systematically in the German TA study.

The HR-TA study illustrates the need of a framework for analysis and evaluation of biosafety controversies. How adequate were the scientific claims represented in the discussions in view of the research goal? The purpose of the TA process was to find an empirical footing for HR biosafety assessment. Thus, we should review and evaluate the way those empirical claims were included. It seems that, in the context of this TA study, the label "empirical" has been one of convenience rather than methodological rigour. An analysis in terms of sufficient SRQs for the purpose can help to shed light on elements in the larger assessment that have been missing. The study is a large-scale test-case for the assessment of genetically engineered organisms in general and therefore may contain lessons for future assessments. In my view, a number of relevant questions have remained unaddressed.

In his introduction to the final report, Van den Daele writes: "In many ways HR technology was and is still viewed as the paradigm of an undesired development conditioned by the prevailing economic and technical trends. That is the very reason why this technology has been made the subject of the TA procedure" (van den Daele 1994: 2). This TA project may indeed be considered to exemplify the larger controversy about biosafety of genetic engineering. The methodological challenge of technology assessment is summarized in one of the reports of the German TA study: "Letztlich geht es um die Entscheidung, wie wir mit Nichtwissen umgehen sollen"[5] (van den Daele *et al.* in: Sukopp and Sukopp 1994: 141).

How well do Van den Daele and his collaborators qualify as 'hazard detectives'? As I see it, the methodological basis of this TA study has been misguided. This will be argued by way of an analysis of the TA process and findings in terms of the alternative methodological basis as presented in Chapter 2. The unsufficing methodology of this TA study is in many ways representative of the larger biosafety debate relating to agricultural biotechnology. On the other hand, this TA study sets an example to future studies in so far as it makes an explicit at-

---

the arguments, counter-arguments, oppositions, second-thoughts, etc. were logically ordered to central claims. The coordinators of the TA process have drawn provisional conclusions".

5 "In final analysis, the decision of how to deal with not-knowing is at stake".

tempt towards a *participative* methodology for technology assessment. One advantage of this explicitness is that it makes it possible for us to recognize the limitations of its methodology.

As an arresting symptom of the methodological incompleteness of this TA study and the frustration it raised with some of its participants, consider the following. The representatives from the environmental groups involved decided to withdraw from the assessment process, by leaving the final conference at which conclusions were to be discussed. They were not satisfied with the TA procedure and its contents. This withdrawal obviously led to bias in scientific expertise and in the outcome of the TA process. The focus should be on the deliberation *process* before we can appreciate the assessment *results* of the TA. The procedure was inappropriate due to a flawed methodology.

One of the complaints was that the project coordinators did not give a fair account of the discussions. Weber (1996), who was one of the experts eventually deciding to leave the TA process, noted that in the expert discussions there were serious differences of opinion about scientific questions such as the relevance of understanding *transposon activity* for the purpose of transgenic HR hazard identification. However, in the final representation some of these controversies had simply disappeared from the scene: "Diesen Dissens ignorierten die Antragsteller in der Auswertung"[6] (Weber 1996: 20). Leaving differences of opinion out of the picture may clearly lead to bias in the results.

What would have been a more rational way of dealing with this lack of consensus? In my interpretation, the differences of opinion in this TA study can be understood as (largely implicit) reflections of *competing alternative sets of relevant questions* (SRQs) which are seen by the respective disputing experts as more useful for the purpose of GEO hazard identification. If we interpret the situation in these terms, then it becomes clear that leaving out dissenting viewpoints will be a serious source of bias. The antagonists have different ideas about what will constitute a sufficient SRQ for the purpose of recognizing possible unwanted effects of transgenic herbicide resistance.

The scientific way to address such a controversy is to find out whether the competing SRQs observe an appropriate level of complexity reduction in relation to the specific problem (detecting possible GEO hazards). Or, in other words, to find out whether the suggested underlying SRQs include relevant questions to deal with the research problem at hand. The issue of contested expertise has not been addressed in this way in the course of the German TA study. The effect of this unsatisfactory approach was phrased as follows by

---

6 "This dissensus was ignored by the TA coordinators in their evaluation".

Weber (1996): "Damit wurde die wissenschaftliche Kontroverse zwar verfahrenstechnisch, aber nicht inhaltlich "geschlossen". Das Potential der verschiedenen, in das Verfahren eingebrachten Ansätze zur wissenschaftlichen Auseinandersetzung mit Nichtwissen wurde nicht genutzt"[7] (Weber 1996: 23).

Van den Daele (1993) has called the departure of the critical participants in the TA process "untimely" and "the wrong signal at the wrong moment" (cf. van den Daele *et al.* 1996: 39), but it is hard to see what else could have been expected. A more constructive approach would have been to interpret differences of opinion as competing SRQs for the purpose of hazard identification. The absence of an analytical framework of this kind has led to the situation that the "green experts" (as they are systematically called by the TA coordinator) found themselves confronted with a "summarizing" SRQ to deal with the problems that was not theirs. From that point onward, they would have been alienated from every conclusion on the basis of this, in their eyes unwarranted, SRQ. The only logical step was, given the chosen TA methodology, to take a distance from the SRQ that had been implicitly molded and thereby biased by the project coordinators. An alternative TA methodology could have led to more constructive results.

If the issue of what would constitute a sufficient SRQ for transgenic HR hazard identification had been central in the debate all along, then participation would have been much easier and it would have been more satisfying for contesting parties to clarify their own position in the larger debate and to make the debate more fruitful. In the following sections, I will argue that the quest for appropriate SRQs has been left implicit in the TA process and that explicit attention for the competing SRQs could have improved the TA process.

— 3.3 Empirical conclusions and relevant questions —

The HR-TA discussions and the experts leaving the final meeting can be reconstructed and understood in terms of the underlying assumed SRQs. The kernel of what critical experts such as Weber and others have done in this TA process, was to suggest the use of a more inclusive SRQ for the purpose of trangenic HR hazard identification (cf. Weber 1995). Van den Daele *et al.* (1996) do present the results of the HR-TA in the form of a "Fragenkomplex" at issue, but the

7 "Although this meant a procedural 'closure' of the scientific controversy, its content was not decided. The potential of the several options for scientific approaches to not-knowing that were brought into the debate, was not used".

relevance of specific research questions to be addressed has not been made the subject of explicit discussion.

Basic scientific methodology implies that we cannot evaluate an empirical claim without interpreting it against its theoretical background. To evaluate the status of a particular empirical claim about biosafety we must therefore also evaluate the methodological status of the background theories from genetics, ecology and evolutionary biology upon which the empirical claim is based. A practical way to do so is to discuss the relevant research questions and the associated burden of proof for inclusion in or exclusion from a sufficient SRQ that results from more general biological background theories. One way in which background theories come to the surface in applied research is in the *definition of the concepts* used.

3.3.1 Defining the phenotype

Consider, as an example, one of the "empirical conclusions" about the possible environmental consequences of HR technology that resulted from the TA investigation: "Das ökologische Verhalten von Pflanzen ist immer anhand der Phänotyps zu beurteilen"[8] (van den Daele *et al.* in Sukopp and Sukopp 1994: 138), and: "Der Phänotype entscheidet"[9] (van den Daele *et al.* 1996: 42). To evaluate this claim, we must refer to its theoretical background. For example, we need to understand the implied definition of 'phenotype'. The "empirical conclusion" is unclear since the term "phenotype" is used for several, different concepts. In this form it seems more like an 'analytical' than an 'empirical' claim, as if the phenotype 'is' the ecological behaviour of a plant.

The phenotype is apparantly understood here as a relatively fixed and stable constellation of traits that determines the ecological behaviour of plants. This suggests that the phenotype itself is *independent* of the environment and undirectionally determines the relationship between a plant and its environment. This definition of phenotype leaves us essentially with one general question to answer for the purpose of hazard identification: what is the phenotype of a GEO? A problem is that this question does not make sense without an analysis of its constituting relevant questions. As Käppeli and Auberson (1997), among others, warn us: "The unpredictability of the phenotype of an organism from

8 "The ecological behaviour of plants can always be evaluated by consideration of the phenotype".

9 "The phenotype is decisive".

its genotype alone, especially its ecological traits and population dynamics, is a major concern for environmental releases of GMOs" (Käppeli and Auberson 1997: 344).

If we would opt for a different definition of 'phenotype', one which includes *interactions* with the environment, then the "empirical conclusion" cited above gets an entirely different meaning. The concept I have in mind takes into account the complex interactions of *genotype*, *phenotype* and *environment* of an organism. The key here would be so-called "norms of reaction" (cf. Gabriel and Lynch 1992; Gabriel 1993). A 'norm of reaction' is defined as: "the function that maps the space of environmental sequences into the space of phenotypic outcomes for a given genotype" (Lewontin 1992: 141). An approach in terms of a trangenic organism and its 'norm of reaction' provide us with an alternative SRQ for GEO hazard identification. This does not imply that we can actually describe the relations *quantitatively* for the purpose of GEO hazard identification, but it enables us to at least acknowledge them in *qualitative* terms. I would opt for the *norms of reaction* approach, which provides room for the appreciation of ecological relationships. If this makes sense methodologically, then the "empirical claim" that "the phenotype is decisive", becomes meaningless. Its plausible ring notwithstanding, it does not represent proper biological science for lack of methodologically articulate concepts.

The problem is related to misguided interpretations of the concept of an 'ecological niche', suggesting that there is a more or less fixed interplay between an organism and its environment. Fincham and Ravetz (1991), for example, have phrased their view as follows: "The ecological lesson is that an organism introduced into a new environment will persist only if it finds a suitable 'niche' – a set of favourable environmental conditions to which it is peculiarly well adapted" (Fincham and Ravetz 1991: 4). However, in modern ecological theory, the concept of 'niche' is mistrusted. As Regal (1985) has argued: "Many ecologists today wisely reject even the use of the word 'niche' in any technical sense. (...) The niche is nothing more than what an organism is and does in any particular part of its range. (...) But then where does this insight leave the evolutionists? If the organism defines itself in the environment, then how can it be the environment that shapes the organism?" (Regal 1985: 15).

As an illustration of the biological complexity involved, consider the following possibly relevant concern. Broer (1995) has pointed out the relevance of asking: "Sind gentechnisch veränderte Pflanzen auch unter Streßbedingungen noch herbizidresistent?"[10] (Broer 1995: 104). It turns out that high temperatures may lead a plant to *inactivate* its transgene expression. For HR crops this would mean that application of the herbicide on hot days could lead to destruction of the crop (Broer 1995: 105). For those depending on the reliability of HR technology this could lead to severe social and economic costs.

This succint analysis of a single "empirical" claim indicates that differences in the implicit or explicit theoretical background of participants in a scientific assessment may profoundly affect the outcome of GEO hazard identification. Explicit theoretical backgrounds and allied methodology are sorely missing in many debates over biosafety. Our prime concern should be to get them out into the open. The claim that "the phenotype is decisive" evades addressing relevant research questions about the role of the environment. As argued in Chapter 2, those who explicitly or implicitly claim that some concern need not be addressed for a particular research purpose, thereby take on the *burden of proof* for justifying this exclusion.

3.3.2 Selective advantage of herbicide resistance?

The "empirical" biosafety claim in relation to the "phenotype" of a GEO analysed above is not an isolated example. The general HR-TA study is riddled with such "empirical claims" that drastically change face when interpreted against an alternative theoretical background. Exploration of the theoretical background and its methodological pitfalls should be a major theme of concern in the process of hazard identification and thus also in the process of technology assessment.

Consider another "empirical conclusion" that resulted from the "scientific discussions" of the TA process: "Das Merkmal der Herbizidresistenz vermittelt jedoch an naturnahen Standorten ohne Herbizideinsatz keinerlei Selektionsvorteile"[11] (van den Daele *et al.* 1996: 42; cf. van den Daele 1994: 120). In this empirical claim, "Selektionsvorteile" (selective advantage) is a central concept. How should we understand the concept in this context? A critical observer may ask, for example: does this empirical conclusion imply that the trait of 'herbicide resistance' will not have a selective advantage in *any* natural ecosystem? A 'selective advantage' typically depends on differences between competitors and their relations to the environment. There was no consensus among the TA experts about this claim: "Herbizidresistenz in Pflanzen stellt grundsätzlich kein ökologisch angepaßtes Merkmal dar"[12] (Weber 1994: 28).

---

10 "Will genetically modified plants also show herbicide-resistance under stressful circumstances?".

11 "The trait of herbicide-resistance does not, however, confer a selective advantage under natural circumstances".

12 "Herbicide-resistance in plants is essentially not an ecologically adapted characteristic".

The fundamental methodological issue should thus be whether the implied interpretation of "selective advantage" is adequate. Does ecological background theory suffice to support an empirical claim that some trait will never have a selective advantage? Or, in other words, which would be the relevant questions we should address to find out whether or not a specific GEO may have a selective advantage or not? As Brandon (1992) has argued: "One needs to know what sort of environmental heterogeneity is at issue if one is to try to measure it, (...) the intuitively appealing idea of measuring some aspect of the external environment will not suffice" (Brandon 1992: 86). Maybe it is true, that herbicide resistance does not give a selective advantage in natural ecosystems, but we can only evaluate this claim if we know what would be a sufficient or useful set of relevant questions to give this claim an empirical basis (cf. Metz and Nap 1997).

It may seem that there is no 'selective advantage' for herbicide resistance. However, even if we cannot detect a 'selective advantage' in natural ecosystems, we are still faced with the selective advantage of herbicide resistance *in an agricultural context*. This is the context in which the difference between *weeds* and *crops* is defined: "A weed can be broadly defined as a plant at the wrong place and/or the wrong time. (...) The weediness of a plant largely depends on the interplay between the intrinsic characters of the plant, in combination with its specific habitat ..." (Metz and Nap 1997: 38).

If we consider the possible transfer of herbicide resistence from a crop species to a weed species, the benefits of biotechnology could soon turn into serious costs. One form of experience with genetic transfer of resistance is the medical context. As Regal (1988) points out: "While we do not know the modes and rates of plasmid transfer in nature, it is well documented that antibiotic resistance transfer can lend an advantage to bacteria in hospitals" (Regal 1988: s37).

The same mechanism may add a new dimension to the 'arms-race' between pests and farmers that has become typical of the use of chemical pesticides in an agricultural context: "... we may inadvertently be creating a 'super weed' when the TG [transgenic] plant or the transgene escapes into wild relatives through hybridization" (Miller 1993: 327). Also, once a weed acquires a firm foothold in an agricultural context, this may be a trigger or 'launch pad' for a more general ecological impact.

An example of the close relationship between crops and weeds in agriculture is given by Kareiva *et al.* (1991): "..., if genes for herbicide- and insect-resistance were inserted into canola (*Brassica napus*), and if these genes were transmitted to neighbouring populations of *Brassica campestris*, it is not hard to imagine wild mustard becoming a much more unmanageable weed than before the release of engineered canola. Of course, crops and wild plants have always hybridized to some extent but never before has there been the possibility of exchanging as many potent single-gene traits as will be available when genetically engineered crops become widespread" (Kareiva *et al.* 1991: 31).

A difference between the magnitude of possible hazards of *trangenic* herbicide resistance as opposed to *conventionally bred* herbicide resistance could result from the character of the specific interventions that genetic engineering has made possible. An important question to address in this assessment is the precise genetic basis of the conferred herbicide resistance. Most genetically engineered forms of herbicide resistant crops will be based on the insertion of single genes: "..., the traits used in genetic engineering are single-locus dominant genes, whereas many of the successes of past plant breeding have been attained with highly polygenic quantitative characters; this is significant because when hybrids are formed and the spread of favourable genes is considered, such spread is much more likely and rapid for a simple dominant gene than for a complex polygenic trait" (Kareiva and Parker 1994: 5).

— 3.4 Burden of proof for comparison —

A central concern in the larger HR technology assessment is the methodological basis for *comparison* between conventional breeding and genetic engineering. The TA coordinator writes: "Entscheidend ist doch, ob man bei gentechnisch veränderten Pflanzen mit anderen Ungewißheiten und unbekannten Risiken konfrontiert ist als bei konventionall gezüchteten Pflanzen"[13] (van den Daele 1994/appendix: 40). What methodology could help to deal with this issue? Does summarizing and ordering the controversial claims bring closer a justifiable answer? In my view, this is a *necessary* but not a *sufficient* prerequisite for rational technology assessment. The same dilemma pervades most, if not all, discussions about biosafety and technology assessment.

In the process of technology assessment it is important to distinguish a *political* burden of proof for making claims on "acceptability" from a *scientific* burden of proof for making claims on "empirical adequacy" or "plausibility". This distinction has not been made in sufficiently clear terms in the German HR-TA. In an earlier publication, Van den Daele (1992) took a skeptical stance on this issue: "On which side the burden of proof for the safety of a technology

---

13 "Decisive is, however, whether genetically engineered plants will confront one with different uncertainties and unknown risks than conventionally bred plants".

should be put is a genuinely political issue, which should be dealt with through procedures of bargaining, compromising and voting rather than by professional analysis. But in fact, the prominence of arguments about the limits of our knowledge and foresight have not put the experts out of business in risk regulation" (van den Daele 1992: 329).

The main reason for the omission to make the distinction between a scientific and a political burden of proof is probably the failure to present an adequate view on the role of science in biosafety assessment and on controversy in applied science. A focus on the relevance of research questions for the purpose of hazard identification opens up a natural perspective on the *scientific* burden of proof that comes with empirical claims. This does not take away the possibility that a *political* decision can be made to 'overrule' the scientific viewpoints. An important source of legitimation for the process of technology assessment is that the science and the politics of choosing or not choosing certain technological options are *distinguished*, even though they cannot be *separated* in practice.

One example of the intimate relation between science and politics, is in the development of biosafety regulation along lines of comparison to the use of conventional organisms. Gassen *et al.* (1991), for example, have argued that: "Ein Zulassungsverfahren für Nutzpflanzen, die mit Methoden der klassischen Züchtung hergestellt wurden, wird nicht durchgeführt und ist auch für die Zukunft nicht zu erwarten. Aus diesem Sachverhalt ist abzuleiten, daß die Gesellschaft bereit ist, das ökologische Risiko bei der Freisetzung solcher Arten zu tragen"[14] (Gassen *et al.* 1991: 70). The complex relationship between the *legitimations of science* and the *legitimations of society* must be given due attention in procedures for assessing the biosafety of genetic engineering (for more detail, see Chapter 6).

3.4.1 Defining 'similarity'

The coordinator of the German HR-TA study has stated that: "The paramount issues that emerged from the assessment discussion were whether the risks of transgenic plants differed from the risks of new plants which have been altered through conventional breeding techniques and how the hypothesis that there

---

14 "A permission process for agricultural plants which have been raised with methods of traditional breeding, will not be imposed and is not to be expected for the future. From this situation can be deduced that society is prepared to take the ecological risk of the introduction of such species".

are special risks for genetically modified plants could possibly be justified" (van den Daele 1994: 16). In an earlier study, Van den Daele already came to the conclusion that, "[o]pinions converged among the experts that genetic engineering was not in fact so 'new' that it could not be compared to traditional biotechnology" (van den Daele 1992: 332).

The strategic importance of these issues for the larger TA study becomes clear from the explicit intention to arrive at: "Normalisierung der erkennbaren Risiken durch Vergleich"[15] (van den Daele 1994: 128; cf. van den Daele *et al.* 1996: 31 and 253). From the point of view of developing biosafety policy and regulation, this is obviously a very pragmatic approach.

The strategy of *comparison* is meant to uncover parallels and similarities between herbicide resistant plants developed through conventional techniques and the HR plants developed with modern recombinant DNA techniques. If it can be shown that no differences in possible hazards exist, then transgenic HR plants should not call for special regulation. This regulatory strategy is also suggested by the US *National Academy of Sciences* and other institutes in the claim that the risks of conventional and new techniques are the "same in kind": "No conceptual distinction exists between genetic modification of plants and microorganisms by classical methods or by molecular techniques that can modify DNA and transfer genes" (NAS 1989: 14). The focus here is on the question: what is the scientific basis of this comparison?

One of the main "empirical conclusions" of the TA study as a whole is: "Es lassen sich für gentechnisch hergestellte HR-Pflanzen keine Risiken zeigen, die nicht auch für HR-Pflanzen gelten, die mit Methoden der konventionellen Züchtung hergestellt werden"[16] (van den Daele 1994: 128). A basic methodological difficulty that must be addressed to evaluate this claim is that, to make a reliable comparison, the research questions considered relevant must be specified in sufficient detail. This means that the participants in the debate should agree about the details needed for the purpose of a particular aspect of comparison, hazard potential in this case. To reach agreement about sufficient relevant detail, the antagonists in the controversy must be explicit about their SRQs. In this reconstruction, the methodological responsibility of the coordinator of a TA study such as discussed here, should be to aim for clarity about alternative possible SRQs as a basis for empirical claims. This important methodological

---

15 "Normalisation of knowable risks by comparison".

16 "One cannot show risks of genetically engineered herbicide-resistant plants, that do not also pertain to herbicide-resistant plants that were made through methods of conventional breeding".

requirement for fruitful discussion has not been satisfied in the German TA study.

One of the methodological issues that is explicitly addressed in the German herbicide resistance technology assessment is the issue of the *burden of proof* for justifying claims. The project coordinator raises the general question: "Who is to bear the burden of proof for claiming that modern genetic techniques and conventional genetic techniques are essentially analogous" (van den Daele *et al.* 1996: 263). In his view, the answer to this question should be: "Bei dieser Beweislastverteilung schlagen die Grenzen unseres Wissens gegen denjenigen aus, der auf Unterschieden besteht"[17] (van den Daele 1994: 135; van den Daele *et al.* 1996: 264). To clarify the importance of distinguishing between a political and a scientific burden of proof, consider the analysis Van den Daele and his colleagues make of the concept of "Gleichheit". In the clarification of their position they write: "Eine Alternative zu dieser Beweislastverteilung scheint es nicht zu geben. (...) Der Begriff der Gleichheit verlöre jedoch jeden Sinn, wenn man dort, wo Unterschiede nich erkennbar sind, bis zum Beweis des Gegenteils von Verschiedenheit ausginge und nicht von Gleichheit"[18] (van den Daele *et al.* 1996: 264; van den Daele 1994: 135).

Politically this may be true, but scientifically it is misguided. This may be illustrated by the more political observation the TA coordinator makes about the burden of proof in relation to technological innovations: "Einer Beweislastregelung, nach der Risikoverdacht nicht von den Gegern der Technik begründet, sondern von den Befürwortern ausgeräumt werden müsste, fallen unterschiedslos alle Innovationen zum Opfer"[19] (van den Daele *et al.* 1996: 50). This may be a practical position in a political context, but it is not an adequate account of scientific methodology for hazard identification. When it is impossible immediately to demonstrate that two entities of comparison are *dissimilar*, this does not imply that they are therefore *similar*. Of course, one could make the *political* decision to accept absence of proof for dissimilarity as sufficient proof for similarity, but then such a decision should not be presented in a

---

17 "In this division of the burden of proof, the boundaries of our knowledge work against those who claim that differences exist".

18 There does not seem to be an alternative to this division of the burden of proof. (...) The concept of 'similarity' would loose all meaning, when, in cases where differences cannot be known, one would assume differences until contrary proof is given instead of similarity".

19 "A division of the burden of proof, in which suspicion of risk need not be supported by the opponents of the technology, but instead be removed by its supporters, would be fatal to all innovations without distinction".

scientific guise. To accept or argue for "Gleichheit" on a scientific basis, we must at least specify the relevant questions that we have considered or will consider to assess the "Gleichheit". If one fails to specify a framework of comparison, then how can the supposed similarity be assessed?

Claims about the "Vergleichbarkeit" of hazard potential cannot be justified without addressing the relevance of research questions about this difference. An essential precondition for identifying possible hazards of transgenic organisms (as well as of conventional organisms) is the developmet of a SRQ that is suited for its purpose. If one should arrive at exactly the same SRQs for both the hazard identification of conventional organisms and for the hazard identification of transgenic organisms, then indeed a scientific basis has been given for the claim that the same biosafety test and regulation suffices for both categories.

It should be noted that even the establishment of this similarity does not imply that the outcome of such a biosafety test will *therefore* be the same for both categories of organisms also. Hazards may be similar in *dimensions* and still be different in *magnitude*. A firecracker and a bomb, for example, both carry the hazard of exploding, but the impact will not be the same in the two cases.

3.4.2 Context-dependence and relevant questions

One of the contested elements of comparison is the relevance of questions about the *context* in which an inserted gene finds a place. One "empirical conclusion" in the final report of the HR-TA study is: "Es gibt weder empirische Anhaltspunkte, noch ein theoretisches Modell dafür, daß bei Transgenen andere oder weitergehende Kontextstörungen möglich sind als bei Transposonen, die natürlicherweise im Pflanzengenom springen"[20] (van den Daele *et al.* 1996: 43). Here again the burden of proof for "Vergleichbarkeit" plays an important role: which are the relevant questions to address for justifying the claim that 'transgenes' and 'transposons' are similar in relation to their disturbances of a genomic context? The TA coordinator applies the same idea of "Gleichheit" here as cited earlier: "Wenn man mit unvorhersehbaren Stoffwechselveränderungen und Kontextstörungen bei allen züchterischen Eingriffen rechnen muß, sind synergistische Risikoeinwände grundsätzlich legitim, aber es fehlt

20 "There are neither empirical reasons, nor a theoretical model, to support that transgenes could cause other or further-going context-disturbances than can be caused by transposons, which jump naturally in the plant genome".

ein Kriterium, um zwischen gentechnisch hergestellten und konventionell gezüchteten Pflanzen zu differenzieren"[21] (van den Daele *et al.* 1996: 261). Does this claim have a sufficient scientific basis?

Tappeser (1994) has criticized, in her contribution to the expert reports, the lack of explicitness about the relevant questions in the debate about the "Vergleichbarkeit" of genetic engineering and conventional breeding techniques. She raises concern about the *genomic stability* of the modified genome as a relevant research question to be included in a useful SRQ for the purpose of giving a scientific basis to the claim of "Vergleichbarkeit" or similarity. In her opinion:

> "Eine genaue Definition dessen, was als Untersuchungsgegenstand betrachtet wird, fehlt allerdings. Ist das gesamte Genkonstrukt Gegenstand der Analyse oder nur der Sequenzanteil der die eigentliche Herbizidresistenz vermittelt? Aus dem Text erschließt sich, daß im wesentlichen nur der Teil des Genkonstruktes betrachtet wird, der die resistenzvermittelnden Eigenschaften kodiert. Eine Begründung, warum diese eingeengte Betrachtung gewählt wurde, wird nicht gegeben. Dies wäre aber wünschenswert gewesen, da durch die Beschränkung eine Reihe von risikorelevanten Fragestellungen nur gestreift werden, die aber für eine Risikobewertung durchaus von Bedeutung sind. Zu nennen sind hier die Sequenz- und Funktionsstabilität z.B. der Antibiotikaresistenzgene und der häufig von pflanzenpathogenen Viren stammenden Promotorregionen sowie deren Transfermöglichkeiten" (Tappeser 1994: 61).[22]

In reconstruction, Tappesers critique can be understood as a plea for applying a more inclusive SRQ for hazard identification as a prerequisite in considering particular empirical claims. Methodologically, claims without a specified SRQ

---

21 "If unpredictable changes in the metabolism or context-disturbances should be considered in all breeding efforts, then synergistic risk-objections are fundamentally legitimate; however, there is no criterium available to distinguish genetically engineered plants and conventionally bred plants".

22 "A precise definition of what is considered to be the research object is missing. Is the complete gene construct subject of analysis or only the part of the genetic sequence which is at the basis of the actual herbicide resistance? From the text can be concluded that essentially only the genetic insertion is considered, which is coding for the resistance-imposing traits. A ground for why this reduced perspective is taken is not given. This would have been desirable though, since, as a consequence of this reduction a number of risk-relevant questions is only touched upon, while they are thoroughly important for a risk evaluation. To be mentioned are the sequential and functional stability, *e.g.*, of genes for antibiotics-resistance and promotor regions which often originate from plant-pathogenic viruses and their potential for transfer".

are not meaningful. Thus, Tappesers comments unearth a serious lack of scientific quality in the TA process: "Eine Reihe von Faktoren, die die Zuverlässigkeit der Expression beeinflussen, werden im vorliegenden Gutachten nach Einschätzung der Kommentatorin nicht ausreichend diskutiert. Zu nennen sind hier die Bedeutung flankierender Sequenzen am Ort der Integration, die Bedeutung unterschiedlicher Promotorregionen, die Bedeutung von Proteinfaktoren und Pflanzenhormonen und deren Interaktion, die die Transkription oder Translation beeinflussen können"[23] (Tappeser 1994: 63). She considers this to be an unwarranted simplification of a sufficient SRQ for the purpose of assessing "Vergleichbarkeit", because: "... genau da, in den Bedingungen, die diese Genkonstrukte zur Etablierung und Expression mitbringen, gibt es gravierende Unterschiede zwischen endogenen Genen und Transgenen"[24] (Tappeser 1994: 65; cf. Gabriel 1993: 109).

Recombinant DNA technology is literally a new technique to manipulate 'contexts': a functional nucleotide sequence or 'gene' can be taken out of the context from one organism and be inserted into another organism. This is a way to change *many context relations* at the same time. In traditional breeding techniques, fewer contexts are changed. Changing the context(s) is the very essence of the appeal of genetic engineering. For instance, in conventional techniques no phylogenetic lineages are crossed, whereas genetic engineering allows even the boundaries between *eukaryotes* and *prokaryotes* to be crossed. This could well make a difference in terms of relevant research questions for studying *context effects.* Those who claim this would not make a difference thereby take on the burden of proof for showing that such additional questions *need not* be considered. If one aims for certain new functions of a GEO, such as the capacity to tolerate the use of specific herbicides, one is thereby engineering context relations that were not open for modification without recombinant DNA technology.

Genetic engineering relies on several context-effects to be in accordance with its objectives. This also explains why genetic engineering attempts very often fail to produce viable transgenic organisms, while traditional breeding efforts mostly produce some viable result even though it may not always be the

---

23 "A host of factors that influence the reliability of the expression, were not sufficiently discussed in the present evaluation in the opinion of the commentator. To be mentioned are the significance of neighbouring sequences at the site of integration, the significance of different promotor regions, the significance of protein factors and plant hormones and their interaction, which can affect transcription or translation".

24 "... just there, in the restrictions for expression which come with these gene constructs, exist serious differences between endogenic genes and transgenes".

desired result. By manipulating the context(s) of genes and of organisms, one takes on the challenge to oversee the relevant considerations for hazard research. The *scientific burden of proof* that is implicated in the claims of a technology assessment or a biosafety assessment can be operationalized for practical purposes in terms of arguments over the *relevance* of individual research questions (see Figure 3.1).

QUESTION:

"What is the relationship between a transgene and its recipient in a GEO?"

Relevant because...	*Not relevant because...*
"Different techniques may create different genomic stability"	"Not different from the relationship between a transposon and a conventional organism"

QUESTION:

"Will transgenic herbicide resistance transfer more easily to weedy relatives than conventionally bred HR?"

Relevant because...	*Not relevant because...*
"Monogenic traits will transfer more easily"	"The 'phenotype' is decisive"

QUESTION:

"What impact will 'naked DNA' (if it survives) have in different environments or 'contexts'?"

Relevant because...	*Not relevant because...*
"Lesser genomic stability of transgenes in GEOs"	"'Contexts' will be changed in conventional breeding also"

FIGURE 3.1: Schematized representation of a procedural operationalization of the scientific burden of proof for inclusion or exclusion of some individual research questions as relevant part of an SRQ for GEO hazard identification (no claim for *completeness* implied).

— 3.5 Molecular biology *versus* ecology? —

Discussion over the *relevance of context relations* has a deep-rooted history in biology. According to Krimsky (1991), the biosafety debate over GEO hazard identification is a reflection of what he calls, "disciplinary fault lines" between the scientific disciplines of molecular biology on the one hand and ecology and evolutionary biology on the other hand. In his view, "[t]he theoretical program of geneticists emphasizes nature's unity and stability, and expresses confidence in human control over biological systems. In contrast, the perspective of ecologists focuses on nature's complexity and interdependence. Ecologists are more comfortable studying nature as a dynamical system comprised of non-linear processes that give expression to nature's indeterminacy" (Krimsky 1991: 135).

Colwell has characterized the supposedly antagonistic nature among disciplines in the life sciences as follows: "One source of disagreement about ecological risk in biotechnology may stem in part from conflicting philosophical views about outliers. Molecular biologists are reassured by the highly deterministic control that is now possible over precise genetic constructions, which in fact may provide a basis for low-risk outliers. Ecologists and evolutionary biologists, in contrast, tend to focus on the inherent complexity and indeterminacy of outcomes in biological communities – the source of 'ecological surprises' that characterize outliers" (Colwell 1988, cit. in NSNE 1997: 37). In my view, the antagonism between more molecular and more ecological approaches is much more *one of history* than *one of principle*. This may be supported by the development of subdisciplines such as 'molecular ecology' and 'ecological genetics' (cf. Daly and Trowell 1996). Both at the molecular *and* ecological level, *questions about biological contexts and environments* are highly relevant for general research purposes and for GEO hazard identification as well (cf. Brandon 1990, Beatty 1982).

A consequence of this difference in research styles or "philosophies" can also be found in the fact that molecular biologists are assumed to provide a more reliable scientific basis than ecologists and evolutionary biologists can offer. The science of ecology has been characterized as being "natural history" rather than a predictive science (cf. Trepl 1987; van der Steen and Kamminga 1991; McIntosh 1985). As McIntosh (1987) has argued: "This boils down to recognition of properties, long familiar to ecologists, namely the complexity, diversity, and difficulty of specification of the objects of study and the lack of regularities in the sense of laws or constants requiring that ecological generalizations be conditional. Pluralistic ecology recognizes that the grail of a unified theoretical ecology is as elusive as the grail of the legend" (McIntosh 1987: 331).

3.5.1 Competing models of biological complexity

The following episode from the biosafety debate may illustrate that such disciplinary controversies are not just a matter of individuals holding unorthodox positions; complete scientific institutes may take opposing positions. This only adds to the seriousness of the problem of resolving such controversies. The German institute for technology assessment (*Büro für Technikfolgen-Abschätzung des Deutschen Bundestages*, TAB in Bonn) has asked two research institutes (the *Institut für Biochemie* in Darmstadt and the *Öko-Institut* in Freiburg) to give their reasoned view on biosafety in relation to the use of recombinant DNA technology (Gassen *et al.* 1991; Bernhardt *et al.* 1991). It is quite interesting and unnerving to see that the two respective scientific institutions presented evaluations with very different implications in relation to biosafety policy.

Kollek (1992) has argued in her evaluation of these two TA-reports that biological processes should be seen in their proper *context*. Her aim in this *Comment* is to explicate the empirical and theoretical foundation of the two positions, to compare them and to identify and specify the correspondences and differences (Kollek 1992: 5). The SRQ analysis underlines the importance of considering *context-dependence* for a proper observation of scientific methodology. The diverging assessments of the two institutes can be traced back to diverging perspectives on what constitutes a sufficient SRQ for the purpose of GEO hazard identification.

The leading idea in the perspective of the Institut für Biochemie, Darmstadt (IBD) is that the hazard potential of a GEO is a *cumulative function* of the respective hazard potentials of the *host organism*, the *vector* that is used to transfer the genetic sequence and the *donor organism*: "Die Beurteilung von in vitro neukombinierten Organismen beruht im wesentlichen auf dem Risikopotential der Empfängerorganismen und der verwendeten Vektoren unter Berücksichtigung der durch das übertragene Gen (DNA-Sequenz) zusätzlich erworbenen Eigenschaften sowie der Spenderorganismen. Somit wird das tatsächliche Gefährdungspotential vor allem von den oben genannten Faktoren abgeleitet. Dieses Verfahren wird als 'additives Modell' bezeichnet"[25] (Gassen *et al.* 1991: 99).

In contrast to this view, the Öko-Institut, Freiburg (ÖIF) has tried to find a more integrated approach, which is characterized by Kollek (1992) as the "syn-

---

25 "The assessment of in vitro recombined organisms is essentially based on the risk potential of the receiving organism and the applied vector with consideration of the additional traits conferred by the transferred gene (DNA sequence) as well as the donor-organism. Thus, the effective hazard potential is predominantly deduced from these factors. This approach is referred to as the 'additive model'".

ergistic model" of hazard identification. In this perspective, it is claimed that we must think of natural systems as essentially dynamic, with the potential for biological traits to add up to more than just the sum of the parts. The two approaches imply different research questions to be relevant as part of a useful SRQ for the purpose of GEO hazard identification.

Broadly speaking, the *additive model* and its assumed SRQ can be seen as emerging from the larger theoretical background of *molecular biology* and the *synergistic* (or *contextualistic*) model and its assumed SRQ as emanating from the disciplinary settings of *ecology* and *evolutionary biology*.

In the rationale of the 'additive model' it is implicitly assumed that enough is known about the donor-organism, the vector and the host-organism to use them as touchstones for an identification of possible hazards. This results in a reductionistic model of the biological mechanisms that are involved with recombinant DNA technology (cf. Chapter 5). The reduction occurs at all three of the related levels of interpretation: the genotype, the phenotype and the environment (cf. Weber 1994).

One (cluster of) research questions that is not specifically addressed in the SRQ underlying the 'additive model' concerns the specific relationships between genotype and phenotype. According to this model any 'addition' or 'subtraction' of genetic sequences will cause a fixed and comprehensible response at the level of the phenotype. This element of *genetic determinism*, however, does not go unchallenged. In fact, to corroborate this statement would require more knowledge than we actually have. Considering these phenomena, Lewontin (1992) notes: "There is, however, only rudimentary knowledge of the causal pathways of development, so the forward mapping of genotypic description into phenotypic description is not possible except in special cases. It is clear that the mapping is not one-one in general, and to the extent that a single genotypic class may correspond to multiple phenotypic classes, a determinate step from offspring genotypes to offspring phenotypes is not possible" (Lewontin 1992: 139).

Those arguing that questions about this level of biological organization need not be raised, thereby take on the burden of proof for justifying this exclusion. Scientific complexity reduction must be legitimated as methodologically admissible, given the specific purpose of research. In relation to *pleiotropy* (the phenomenon where a single gene is responsible for a number of distinct and seemingly unrelated phenotypic effects) and *epistasis* (the phenomenon where one gene modifies the expression of another gene that is not an allele of the first), Krimsky (1996) remarks: "A new generation of questions are beginning to be asked about genetically modified organisms. Without direct empirical knowledge, some scientists are resorting to historical, theoretical, or analogical reasoning to address this new generation of questions" (Krimsky 1996: 247).

In the omission of the IBD to consider such questions as possibly relevant, we may see a reflection of the history and genesis of molecular biology and its strong early relationship to physics. As Keller (1990) has argued, this historical background is not a neutral element of molecular biology as we know it today. The rise of molecular biology – of what Kay (1993) has called the "molecular vision of life" – has brought about a transformation in biology, "of language, of focus, of methodology, of the very definition of what constitutes either a legitimate question or an adequate answer" (Keller 1990: 406). The effect of this larger transformation seems to be reflected in the specific SRQ that is put forward by the *Institut für Biochemie.*

Keller (1990) has phrased the ensuing threat to our thinking and questioning in relation to biological complexity in her view on the history of molecular biology: "Other processes, less identifiable, and less controllable, are bracketed in the double name of intellectual economy and technological efficacy. In this way, the very meaning of knowledge – what counts as knowledge – is shaped by a tacit instrumental mandate, perhaps even when that mandate has been forgotten" (Keller 1990: 408).

The basic assessment that underlies the 'synergistic model', on the contrary, is the idea that recombinant DNA technology is unprecedented. According to the ÖIF this new technology is capable of disturbing and renewing relationships in nature in a way and to an extent that essentially transcends ("die wesentlich über das hinausgeht"), what has been observed and achieved in natural processes up to now. Genetic engineering can be characterized, according to ÖIF, as an "injuring of contextual relationships" ("Verletzung von Kontextbezügen"). These fundamental disturbances, the ÖIF contends, have the potential of effecting unpredictable and irreversible ecological changes with possibly catastrophic dimensions (Kollek 1992: 8; cf. Bernhardt *et al.* 1991).

It is argued by the ÖIF that a conjuction of known biological elements can give rise to different effects than a plain addition of the seperate effects of these elements would lead one to suspect (= synergy). According to the IBD, synergistic effects can be considered as outliers ("Einzelfällen") upon which it is not realistic to base our risk-assessments, as it is done in the 'synergistic model' (Kollek 1992: 10). To the ÖIF, on the contrary, this argument is unacceptable, since special cases of natural relationships such as synergy, in their interpretation, are part of 'natural history' and an indication of unpredictable complexity in natural communities.

The only way to resolve this opposition about the relevance of "synergy", is to specify the relevant research questions needed to interpret the meaning of "Einzelfällen" or "outliers" and "natural history" in this context. This approach shows that there is *no inherent antagonism* between the more molecular and the more ecological life sciences. Molecular biology and ecology must *both* contribute to useful SRQs for the purpose of GEO hazard identification.

One example of the *complementary* perspectives of molecular biologists and ecologists is the different *time scales* they study. Molecular processes will typically develop in split seconds, while ecological and evolutionary processes often develop on a scale of many years. This shows how the "disciplinary fault lines" cannot demarcate a choice of *either-or*, but rather should imply a decision of *and-and*. Williamson (1992) has expressed the necessity of an inclusive rather than an exclusive choice of relevant questions as follows: "Molecular ecology is an essential ingredient in ensuring that risks are assessed efficiently" (Williamson 1992: 3).

— 3.6 Participative technology assessment? —

The structure of this TA debate and its lack of explicit attention for general methodology and the relevance of research questions for the purpose of hazard identification, is in many ways typical for other ongoing biosafety controversies. Many critical participants in the debate make good points, but in most cases assess the faulty methodology only implicitly. This leaves room for avoidable mistakes in the future. The German TA study has been a positive example in the sense that it explicitly touches upon many of the relevant scientific problems involved.

In the end, the discussion was not fundamental enough to deal with the important question that was raised by the project coordinator in his general reflections: "Daß es in all diesen Fällen Ungewißheiten und Unsicherheiten gibt, die nicht auszuräumen sind, wurde von allen Verfahrensbeteiligten zugestanden. Umstritten war allein, wie mit solchen Ungewißheiten und Unsicherheiten umzugehen ist. Sind sie ein hinreichender Grund, auf die Einführung trangener HR-Pflanzen zu verzichten?"[26] (van den Daele 1994: 128). The latter question can only be answered in a political arena on the basis of the best available scientific assessments of our knowledge and its limitations.

The more political analysis of the coordinator of this HR-TA may be reflected in his concern that: "Not to implement a new technology does not restore an idyllic state of unspoiled nature but leaves us with an old technology which is

26 "All parties in the process agreed that there are unknowns and uncertainties in all these cases, that cannot be removed. The controversy was only about a proper way to deal with those unknowns and uncertainties. Is this a sufficient ground to abstain from the application of herbicide-resistant plants?"

not necessarily better understood or less risky" (van den Daele 1992: 330). In my view, a truly *participative* technology assessment should find ways to incorporate critical or "green" experts as well to incorporate the merits of public participation.

The biased outcome of the debate was the unavoidable result of a faulty TA methodology. Summarizing the claims and positions is not enough to stimulate an informed debate. It is telling that none of the participants in the TA study came to change his or her mind about a subject and that the TA coordinator writes about the involved experts as being "opponents" and "promotors" of the technology under assessment (van den Daele 1993). This does not seem to set an open-minded stage for a constructive and participative technology assessment.

Incorporating the efforts of critical parties is a *conditio sine qua non* for a participative technology assessment. One of the worries among the so-called 'green experts' was the limited funding they had to sustain their research efforts. In future assessments this practical *and* theoretical problem should raise concern about the scientific bias that may be the result when critical voices are silenced by a lack of resources (cf. Weber 1996).

The subtitle of the final report of the HR-TA study is: "Modell einer partizipativen Technikfolgenabschätzung zum Einsatz transgener herbizidresistenten Pflanzen."[27] In my view, this "political experiment" falls short of qualifying as a *model* or *example* for future assessments. As a consequence of the limits exposed in this chapter, the "participative" character of the larger project was obstructed. The participating experts who walked out did not consider the process as sufficiently participatory in its methodological procedure. It is essential in a TA study to acknowledge the need for *scientific* justification of complexity reduction before making *political* decisions (for more detail, see Chapter 6). This can be done by concentrating on the relevance of research questions and the associated burden of proof for complexity reduction.

27 "Model of a participatory technology assessment for the application of transgenic herbicide-resistant plants".

CHAPTER 4

Artificial Controversies over Hazards of GEO Release

— 4.1 Evaluating contested biosafety claims —

In the present chapter, the methodological framework developed in Chapter 2 is applied to some of the central controversies in the biosafety debate. Interpreting biosafety controversies as discussions about sufficient SRQs for GEO hazard identification, the scientific burden of proof for including and excluding specific research questions as (ir)relevant for consideration is used as a methodological criterium to evaluate empirical claims on biosafety. For every individual contested question concern may be raised about whether it should be part of a useful SRQ for the purpose of identifying possible hazards of GEO release, or not.

A biosafety claim includes an implicit statement about a sufficient SRQ for hazard identification. Starting with a specific claim, it may be asked to what extent it is supported with sufficient scientific backing. The present research is concerned with the relationship between claims and their backing. For example, as demonstrated in Section 2.3.3, to make a claim about the *competitiveness* of an organism without being prepared to explicate the relevant questions about the *environment* of the organism, means failing to give scientific legitimation to this claim.

Methodological omissions like this frequently occur in biosafety debates. This leaves politicians without a proper scientific basis for biosafety policy. As a remedy to this kind of flawed scientific methodology, the SRQ approach is applied to long-standing scientific controversies over the biosafety of the environmental release of transgenic organisms. The list of controversial claims analysed below is not exhaustive. There are more biosafety controversies to be considered, but they can all be reconstructed and evaluated along the same lines of analysis.

A number of claims and counter-claims in the biosafety controversy keeps being repeated without a proper scientific basis. To move beyond this stalemate, an analysis of the methodological basis of particular claims is required. If the relevant research questions of a particular claim cannot be specified and justified by biological theory, this implies that the burden of proof for that claim should be handed back to the expert(s) who put forward the claim in the first place. The kernel of biosafety controversies can be reconstructed and understood as dealing with the *choice and selection of relevant research questions.*

— 4.2 Product *versus* Process —

In recent publications in *Science, Nature, Bio/Technology* and *Trends in Biotechnology* (Miller 1993; Miller and Gunary 1994; see also Miller 1997), Miller *cum suis* criticize what they call a "horizontal framework" in the biosafety regulation of deliberate release. According to defenders of the "horizontal framework" there is, "something systematically similar and functionally important about the set of organisms whose only common characteristic is their manipulation with the techniques of the new biotechnology, and (...), therefore, scanning across various organisms or experiments that use recombinant DNA techniques constitutes a useful category" (Miller and Gunary 1993: 1500). Miller argues that there is good reason to think that this perspective is misguided: "International organizations and professional groups have explored repeatedly the question of whether there are incremental risks of applying 'the new biotechnology'. Sometimes dubbed the 'product versus process controversy', the basic question has been whether the use of the techniques of recombinant DNA technology would require new paradigms – in governmental regulation, in planning and performing risk assessment, and even in the public psyche. For some time, this question has been resolved in the negative" (Miller 1994: 292).

Miller (1994) cites the following justification for his position from a "joint statement", made by several scientific committees: "The properties of the introduced organisms and its target environment are the key factors in the assessment of risk. Such factors as the demographic characterization of the introduced organisms; genetic stability, including the potential for horizontal transfer or outcrossing with weedy species; and the fit of the species to the physical and biological environment. (...) These considerations apply equally to both modified or unmodified organisms; and, in the case of modified organisms, they apply independently of the techniques used to achieve modification; that is, it is the organism itself, and not how it was constructed, that is important" (cit. in Miller 1994: 293).

Although one may agree with this general statement, it does not imply much for biosafety regulation. Perhaps the same "considerations" applied to modern and to traditional techniques would lead to the conclusion that a particular modern technique generates products with hazards of a particular type that surpass hazards associated with any older technique. Surely such a technique should call for additional regulation. The fact that the same *general considerations* apply to two categories by no means implies that the *specific outcome* of the considerations will therefore be the same also. The only way to clear this up is to specify the questions that are relevant for making the difference.

4.2.1 Regulating the process or the product?

In an attempt to downplay the importance of this controversy, Miller (1994) claims that he has the scientific community largely at his side in this issue: "The notion that there is something systematically similar and functionally important about the set of organisms whose only common characteristic is genetic manipulation with the techniques of the new biotechnology (a 'horizontal approach') therefore contradicts the currently accepted consensus view" (Miller 1994: 292). According to Miller and Gunary (1993), there is, "... a wide consensus that risk is primarily a function of the characteristics of a product (whether it is inert or a living organism) rather than the use of genetic modification" (Miller and Gunary 1993: 1500). Miller (1994) sees only one exception to this general agreement: "The single major exception to this scientific consensus appears to be views held by a part of the ecology community, which insists that rDNA technology and its products present additional or differential risks" (Miller 1994: 294).

As a more viable alternative, Miller suggests we should adopt what he calls a "vertical framework" of biosafety regulation, which is more compatible with the "consensus" view. The prevailing consensus, according to Miller, "is based less on empirical data and more on extrapolation from general scientific principles, especially those derived from the accumulated knowledge of biological principles and from our understanding of evolutionary biology" (Miller 1994: 292). In the "vertical framework" that Miller proposes, "the most rational approach to risk-assessment when risk is not readily demonstrable (a situation sometimes referred to as 'very low risk') is to use established scientific principles and to identify significant gaps in understanding that can be addressed by the conduct of properly designed experiments. This approach would rely heavily on previous knowledge about the behaviour of genetic variants of organisms that have been manipulated by conventional methods or that are present in nature, under various conditions of testing and use" (Miller 1994: 295). In my view, the only way to evaluate this type of claims is by specifying them in terms of the relevant research questions necessary to give them a scientific basis.

Miller, who is a former official of the *Office of Biotechnology* of the US *Food and Drug Administration* and is now affiliated to the *Hoover Institution*, is indeed not alone in his perception of adequate biosafety regulation. The same stance is taken by Davis (1989), who has argued that GEO release is being "overregulated", because it is based on assumptions that are contradicted by "evolutionary principles" (Davis 1989: 864).

How appealing is this "vertical" remedy to the "horizontal" approach from a scientific point of view? Is the conclusion that "we know enough" warranted in relation to the biosafety assessment of deliberate GEO release? These questions raise the problem of the methodological status of genetics, evolutionary biology and ecology, as three of the most relevant domains of background theory, in the context of applied biotechnology research.

The regulatory debate is burdened by ambiguous concepts. The contrast between a 'process-based' and a 'product-based' regulation is not so clear as Miller and Gunary want us to believe (cf. Tait and Levidow 1992). In fact, *both* parties need to address questions about the *relationship* between processes and products. The labels 'product approach' and 'process approach' are misguided and they generate artificial controversy.

4.2.2 'Precision' of genetic engineering

The *Organization for Economic Cooperation and Development* (OECD 1993) has recognized differences between the processes of conventional breeding and genetic engineering: "..., the conventional and newer molecular techniques differ in two respects. Molecular techniques allow firstly a greater diversity of genes to be introduced into organisms and, secondly, in general, greater precision in the introduction of genetic material, yielding a more thoroughly characterised and potentially more predictable organism. As the characteristics of the organism depend on its genetic make-up, the view has been expressed that there may be a particular concern if there is a lack of experience with organisms having particular genetic combinations from different sources" (OECD 1993: 7).

At the same time, the OECD acknowledges that it is not quite clear what a proper interpretation of "precision" would be in this context: "In contrast to conventional plant breeding, which is a highly random process, it is possible with molecular techniques to insert the precise genetic information of interest (...) In general, a particular gene will encode the same primary product regardless of the organism in which the trait is expressed or from which it originated. For this reason, the exact primary gene products can more likely be predicted when a plant is modified by molecular techniques than when undefined or uncharacterised genetic material is introduced through less precise methods such as cell fusion or chromosome substitution. However, even though the approximate phenotype usually can be predicted, it is not entirely possible to predict exactly the interactions of a gene in a new genetic background. As a consequence it is necessary to conduct a programme of field tests in the same way as for lines developed by conventional techniques" (OECD 1993: 15). Some biosafety experts take the *precision claim* a step further: "Because plant genetic engineering is based on the transfer of defined genes, the risks are predictable and can therefore be controlled" (Schell 1994: 17).

To validate this claim we must specify what research questions could give a basis to the claim that the new techniques of genetic engineering would be more "precise" than the conventional techniques of traditional breeding and cell fusion. One methodological prerequisite to validate the precision claim is that we must give a more specific definition of what a 'gene' (and a 'transgene') is: "The better one understands the function of the DNA that is being introduced, the better one will be able to predict the consequences" (NSNE 1997: 47; cf. Hubbard 1995).

An important limitation to traditional breeding is that it can only work with 'Mendelian' traits or polymorphisms. These are traits that can be selected for (Regal 1994: 8). All mammals have two eyes, for example, without variation. Therefore we will never be able to breed a variety with four or three eyes, because it is simply not a trait that is open to selection *for lack of variation*. Since rDNA techniques are not restricted to the variation that comes naturally, this new avenue opens up biological options that are principally missing in traditional breeding techniques.

One research question that should be addressed to validate the precision claim is: are there differences between a conventionally bred genome and a genome that is the product of genetic engineering? In the analysis by Regal (1994): "Breeders have selected for phenotypic traits over the last several thousand years, and it is not completely clear what the underlying changes in genomes have been" (Regal 1994: 7). Traditional breeding proceeds by selecting between the genetic options (the alleles) that are already present in the organism.

Selecting between available alleles is not the same as liberally choosing new genes, however, and therefore we are not warranted to consider traditional breeding as the long-tested example or model of recombinant DNA technology. The genetically modified organisms that are the product of genetic engineering are not the result of selecting between different alleles, but are effected by inserting, deleting or moving around genes, giving us, "access to the non-Mendelian portions of the genome" (Regal 1994: 8).

An important problem in making any realistic assessment of the difference between traditional breeding and genetic engineering is that a genome is a *dynamic system* in itself (cf. Brandon 1992). Consider, for example, the following observations by Weber (1994): "Die artfremden oder synthetischen übertragenen Pomotor- und Strukturgensequenzen können im Empfängergenom keine "passende" Integrationsstelle finden. Sie werden an beliebigen Stellen im Genom integriert. Dies geschieht häufig mit mehreren Kopien des fremden Genkonstrukts gleichzeitig. (...) Bei natürlichen oder im Rahmen konventioneller Züchtung herbeigeführten Kreuzungen befinden sich die neueingekreuzten Allele im allgemeinen an einer bestimmten, für das betreffende Gen charakteristischen Stelle auf dem Chromosom (Genlocus). Das heißt die serielle Anordnung der Gene auf den Chromosomen bleibt erhalten"[28] (Weber 1994: 23).

This dynamic picture of the genome raises questions about "genetic stability" of the transgenic genome, a concern which is also mentioned as relevant by the *Ecological Society of America* (Tiedje *et al.* 1989). Since this is one of the factors that may influence the establishment of a gene and its product in an ecosystem, it may profoundly affect the possible effects of introduction. However, it is given limited attention: "There is an implicit presupposition that genetic modifications are stable. Considerations on how the genetically modified organisms will behave *in vivo* and in the long run are fully omitted" (NSNE 1997: 57).

Genetic stability of a GEO may be an important biosafety concern. Jäger and Tappeser (1996) suggest additional research questions should be considered relevant, since: "... risk assessment cannot be stopped with the assumption that a given GEO will not survive. Rather it should be extended to the fate of its DNA which may be stably integrated, eventually expressed and by chance may provide advantage to indigenous microorganisms" (Jäger and Tappeser 1996: 66). This DNA survival may not be without consequences: "... the inserted DNA (the 'transgene') may be toxic to humans or animals; the second concern is that the transgene will 'escape' from the confines of agriculture and lead to undesirable environmental change" (Raybould and Gray 1993: 200).

---

28 "The foreign or synthetic transferred promotor and structure sequences cannot find a "fitting" integration site in the recipient organism. They will be integrated at contingent locations. This often occurs in a number of copies of the foreign gene construct at one time (...) In naturally or in the context of conventional breeding realised crosses, the newly integrated alleles will generally be located at chromosomal sites (gene locus) which are characteristic for the specific gene. This means that the serial structure is preserved".

In considering these questions as possibly relevant, the following observations by Ho (1997) should also concern us: "DNA can remain in the environment where they can be picked up by bacteria and incorporated into their genome. DNA is, in fact, one of the toughest molecules. Biochemists jumped with joy when they didn't have to work with proteins anymore, which lose their activity very readily. By contrast, DNA survives rigorous boiling, so when they approve processed food on grounds that they do not contain DNA, ask exactly how the processing is done, and whether the appropriate tests for the presence of DNA have been carried out" (Ho 1997: 6). Others have raised concerns about possible GEO hazards of "genetic pollution" (cf. van den Daele *et al.* 1996: 249) and the "invasiveness of transgenic genes" (Kareiva *et al.* 1991: 37; Mergeay *et al.* 1994: 70).

4.2.3 Questions about the role of transposons

The fluidity of genomes may prove to be a hazard factor in the release of GEOs (cf. Wheale and McNally 1988; Goodwin 1996), and thus may be a relevant concern to incorporate in a SRQ for GEO hazard identification. Possibly relevant concerns include so-called 'mobile genetic elements' or 'transposons', segments of DNA that are capable of moving to new locations in the genome. *Should questions about change in the genome produced by transposable genetic elements be included in a SRQ for hazard identification?* We need a characterisation of transposons and the role they play in the larger genome to evaluate this concern.

Different views on the possible impact of transposons may result in different conclusions about the possible hazards of applying GEOs. Some have argued in relation to the role of transposons that, "... their behaviour is essentially the same as that of foreign DNA transferred to target organisms in gene technology" (Wöhrmann *et al.* 1996: 2). This would imply that *transgenes* and *transposons* should be understood along the same lines of interpretation and that the same relevant research questions apply.

As yet, no consensus exists about the character of transposons. One interpretation is that transposons are a form of parasitic DNA that reduces the 'fitness' of the genome in which they reside. Another interpretation holds, on the contrary, that transposons improve the fitness of their host (Condit 1991). Choosing one or the other interpretation as a basis for a SRQ for the purpose of GEO hazard identification may thus make an important difference.

The dynamics chosen as the basis of an interpretation may indeed have a profound effect on practical choices. Genes that are introduced in a GEO might be mobilized into transposons, and from there onto plasmids. Thus, they may spread into new organisms. The *population dynamics* of transposons could be relevant to identify the possible hazards of recombinant DNA technology.

Condit (1991) has made a comparison of the two interpretations for bacterial populations and their transposable elements. If a high number of transposons in a bacterial population has *detrimental effects* on its fitness, then this will reduce the likelihood of genes inserted through genetic modification being transferred to other populations. However, if the inserted genes carry a fitness advantage for the receiving organism, this may overcome the decrease in fitness effected by transposon copy number. Starting from the alternative interpretation that transposons may in fact *increase* the fitness of an organism by *enhancing its mutation rate*, this will stimulate the spread of inserted genes (with or without fitness advantages).

Can we argue that questions about the role of transposons need not be included in a SRQ for GEO hazard identification? If not, such questions should be included. As yet we do not know enough about the character and behaviour of transposons to claim understanding of their role in the genetically engineered organism. In an assessment of possible impacts of transposons in genetically engineered organisms, Sayre and Miller (1991) concluded: "It is clear from this review that very little is actually known about the ways that genetic material is transferred in the environment and with what frequencies. There are numerous questions that must be addressed in determining the potential and actual risks associated with the release of MGEs (mobile genetic elements) containing genetically engineered DNA sequences into the environment" (Sayre and Miller 1991: 167). In my analysis, those claiming that such research questions need not be addressed for the purpose of GEO hazard identification (given the required argument for their relevance), thereby take on the burden of proof for justifying this claim (see Figure 4.1).

The 'process versus product' controversy is *artificial*, because it results from a lack of specified research questions. A more articulated discussion about a sufficient SRQ for the purpose of GEO hazard identification will require us to consider research questions about process *and* product. Those arguing that the specifics of the process(es) of genetic engineering need not be considered, thereby take on the burden of proof for excluding the related research questions from a sufficient SRQ as not relevant for the purpose of GEO hazard identification. A dispute about the *relevance* of these questions for the specified research purpose would constitute a *fundamental* scientific controversy.

QUESTION:

What is the *genomic stability* of transgenic inserts compared to conventional organisms?

Relevant because...	*Not relevant because...*
"Less stable integration of transgene(s) may increase hazard of transgene transfer"	"*Transgenes* can be compared to *transposons* and therefore pose no additional hazard"

QUESTION:

What is the difference between genetic variation available to *transgenic* and *conventional* changes?

Relevant because...	*Not relevant because...*
"Traditional breeding uses existing variation; genetic engineering introduces *new* variation"	"The *phenotype* is decisive"

QUESTION:

What is the *fate of transgenes* or 'naked DNA' outside the GEO?

Relevant because...	*Not relevant because...*
"Transgenes can confer hazards in new contexts"	"Horizontal gene transfer is widespread in nature and will not pose new hazards"

FIGURE 4.1: Schematized representation of the scientific burden of proof for inclusion or exclusion of specific research questions as relevant or not relevant part of an SRQ for the hazard identification of a transgenic organism (no claim for *completeness* implied).

— 4.3 Case-by-case *versus* Generic review —

Consider another argument of Miller in his campaign against existing regulation. He claims that we should strive for more *generic* forms of regulation in contrast to *case-by-case* review (Miller 1987, 1994a, 1994b, 1995). Here again, Miller argues that his claim is based on "scientific principles". Below, an analysis is given of this controversy and the misunderstanding that causes it. Core of the presented evaluative approach is that the two regulatory philosophies *both* depend on a specification of the relevant research questions. If this perspective is taken, the two types of proposed review are not necessarily in conflict.

On the basis of his general claim that the new technology poses no additional risks, Miller (1994) disapprovingly notes that: "... the *Ecological Society of America* (ESA) has called for the 'case by case every case' governmental review of all field trials with recombinant organisms. If extensive review processes, rather than an individual investigator's assessment, are required in every case, this implies ... a substantial bureaucracy [which is unacceptable]" (Miller 1994: 294). Davis (1987, 1989), has also argued that GEO release is being "overregulated", because the regulation is based on assumptions that are contradicted by "evolutionary principles".

One "evolutionary principle" mentioned by Davis (1989) is, "... that evolution selects for balanced, harmonious genomes, and so any alteration that humans introduce into an organism found in nature is almost certain to decrease its competitiveness in the natural world" (Davis 1989). This principle is plainly false. Various examples of genetic manipulation enhancing competitiveness are known (for references see Tiedje *et al.* 1989: 302-303). Besides, it is important to note that it is not only fitness in the form of competitiveness that determines the establishment of a new species. Other evolutionary forces such as genetic drift can make the effect of 'fitness' less important (cf. Gliddon 1994: 43).

At the background of Davis' "evolutionary principle" lurks the idea that evolution leads to perfect adaptation, and maximal fitness. That idea must be resisted. Concepts of perfect adaptation and maximal fitness are meaningless. Adaptation and fitness must be conceptualized, for example, relative to particular environments. Further, adaptation is constrained by genetic variants that happen to be available in nature, and organisms cannot reach all "adaptive peaks" because peaks are separated by valleys. Thus, genomes are not balanced and harmonious in general. If Davis' line of reasoning were cogent, no evolution would be possible at all.

An important challenge to reasonable GEO hazard identification is to find relevant *categories* that enable us to deal with cases of deliberate release. An evaluation of every individual case without the use of generic categories is not a rational option. In that respect the controversy over case-by-case *or* generic review is artificial and the consequence of conceptual confusion. The real problem is that no system of relevant categories is now firmly in place, as yet.

Potential hazards attach to many different features of modified organisms, so we have categories along all sorts of dimensions. Few individual cases will belong to the same categories, and in practice this could now lead to an assessment of every case individually. Our aim should be to arrive at generalizations about salient dimensions that permit us to disregard others. For example, *if* we would know that organisms with a low fitness cannot have undesirable features such as pathogenicity or weediness, we could exempt a major category (organisms with a low fitness) from regulation. Unfortunately, proper generalizations are hard to come by. The one about fitness, for example, is problematic since fitness is not a unitary feature; fitnesses depend on the environment (for more detail, see Section 4.3.2). Also, organisms with a low fitness may get involved in lateral gene transfer.

The dispute over the case-by-case approach is in fact a dispute about *candidate-generalities* or useful SRQs that are needed to justify exemption from regulation for particular categories of organisms. Justification will have to rely on biological theory. Therefore, we must know what levels of generality can be attained in biological theorizing. In fact highly general theories or models are possible only to a limited extent in biology (cf. van der Steen and Kamminga 1991). The challenge for biosafety assessment is aptly characterized by Van Elsas (1995): "In the initial stages of assessment, a case-by-case approach is also advisable, judging each case on its own merits. Subsequently, "case" in the case-by-case approach may stand for larger categories of cases which are grouped together based on logic and common sense, e.g. similarity of characteristics of the organism or insert" (van Elsas 1995: 125). This also indicates that between *case-by-case review* and a more *generic review* there is less of a dichotomy than the controversy over it would suggest.

Regulatory practice has already been developing towards more generic review for field tests of transgenic plants: "Initially, these were reviewed on a case-by-case approach. However, in 1993, the USDA began relaxing its requirements for field-testing transgenic crops. New rules allow companies to field-test six common crops that are genetically modified (corn, cotton, tomatoes, potatoes, tobacco, and soybeans)" (Krimsky 1996: 250). Apart from exemption of specific transgenic *species*, exemptions are also developed for specific *changes* such as gene deletions (cf. EPA 1994), and for specific *traits* of a GEO: "Full legislative clearance of this trait [kanamycin resistance, used as selection marker] is therefore acceptable" (Nap *et al.* 1992: 239).

The difficulty of finding scientifically based generic exemptions for regulation becomes the more apparent if we consider the *global scale* of the concerns that need to be addressed: "Regulation of the direct ecological risks is scientifically justified – these need to be considered in a global context because local evaluation of risk does not suffice to ensure global safety" (Kareiva and Parker 1994: 3).

4.3.1 Natural history and adaptation

It has been argued that the *generic* biosafety claim that GEOs will be "crippled" by an "excess baggage" (an additional metabolic load or burden) cannot be decided and that the controversy over this issue must be seen as one of opposing "plausible" claims (cf. von Schomberg 1997). In my view, this is another example of a seemingly fundamental controversy which can be dismissed by proper methodological analysis. Generic biosafety claims of this kind are *unscientific* because they are based on a too simplified and thereby biased SRQ. A more balanced account is given by Tiedje *et al.* (1989): "Many engineered organisms will probably be less fit than the parent organism, although some important exceptions may arise" (Tiedje *et al.* 1989: 298).

Regal (1985, 1986, 1987, 1988, 1990, 1993, 1994) has argued that there is no scientific "plausibility" in the generalized "excess baggage" claim. His arguments can be understood as a call for a more inclusive set of relevant questions for the study of GEO hazards. Those claiming that some of these questions need not be included in a SRQ for this purpose, thereby take on the burden of proof for justifying this claim.

Regal (1985) has criticized *generic safety claims* put forward in biosafety controversies by exposing the lack of biological realism of the "balance-of-nature paradigm" upon which such claims are based. He argues that there is an, "oversimplification of ecological issues in the claims that it will be quite safe to release any genetically engineered form into the environment" (Regal 1985: 13). Proponents of generic biosafety claims assume that *any* alterations will reduce the adaptive potential of the organism because they believe that, "organismal forms are so highly perfected or optimized by millions of years of evolution that any alteration of them will make them less effective in nature unless they are tended by humans" (Regal 1985). The logical conclusion from this line of thought is that, "the chances of [GEOs] causing any damage are nil" (Regal 1985: 13).

It may well be that a genetically engineered trait will present a cost (*e.g.*, a metabolic burden) to the transgenic organism, but the trait could also have benefits for a GEO in terms of a selective advantage. As Regal (1988) has argued: "The issue must be the ratio of costs and benefits" (Regal 1988: s36). Consider also the line of reasoning suggested by Sharples (1987): "The fundamental premise of evolutionary theory is that natural selection, the dominant force responsible for adaptations of organisms to their environments, operates on generic alterations or novelties – mutations, rearrangements, and acquired accessory elements, such as plasmids – to produce evolutionary change. It follows that at least some genetic alterations improve the abilities of organisms to survive, reproduce, compete for resources, or invade new habitats" (Sharples 1987:

1330). The consequence of this line of thought is that, "..., the extent to which a handicap actually does reduce fitness can depend on the environmental context in which an organism finds itself. Suppose, for example, that there is an ecological system with two resources and one consumer organism. This organism efficiently uses one resource and leaves the other unexploited. Suppose, then, that a mutant form of the organism arises that is marginally equipped to use either resource. It cannot compete with its efficient parent for the resource the parent favors, but since the second resource is also available, the mutant has it all to itself. Despite its genetic handicap, the mutant form may be sufficiently fit to survive on the unexploited resource" (Sharples 1987: 1330).

Regal and Sharples criticize the relevance of the research questions upon which *generic safety* arguments and *balance of nature* reasoning are based. A focus on the relevance of individual research questions helps us avoid the relativistic overtones of the notion of a "paradigm". An important problem of biosafety controversies is not so much the *incommensurability* of viewpoints (cf. Kuhn 1963), and therefore the impossibility of comparing or evaluating them, which is an essential element of an interpretation in terms of conflicting paradigms. A SRQ can be qualified in terms of its *relevance for a specific research purpose* such as biotechnology hazard identification.

Regal (1985) proceeds to present an alternative and, as he claims, a more useful SRQ for the purpose of biosafety assessment, which he calls the "individualistic paradigm". In terms of an SRQ analysis, what this alternative approach does to the process of hazard identification is that it includes additional aspects of biological complexity as relevant for concern and thereby additional possibilities of detecting unwanted effects. The net effect of applying one or the other SRQ is that the "balance of nature" approach will *overlook* concerns that will be made *visible* by an "individualistic" approach. This implies that the latter will be a more useful methodological basis for GEO hazard identification.

Regal (1985) argues that concerns relevant to biosafety assessment which are included in the "individualistic" SRQ and excluded from a "balance of nature" SRQ are in large part a result of the definition of "adaptation" that is applied: "Adaptation can be one of the most confusing terms in biology (...) Adaptation could be defined on the one hand so that almost no organic features are adaptive or on the other hand so that almost all organic features are adaptive" (Regal 1985: 17).

Adding these concerns to our picture of ecological complexity, we arrive at drastically different expectations about the possibility of GEOs to survive in natural circumstances. Choosing one or the other SRQ as a basis for interpreting biosafety issues can be compared to deciding to picture a stable state in nature, "as a house of cards or as one position of a gyroscope that under the right circumstances can be shifted to a new position" (Regal 1985: 16). Relying on one or the other mental picture of biological complexity will result in different outcomes of GEO hazard identification.

Focusing the larger biosafety debate on discussions over relevant research questions will be the best methodological basis for progress in the respective disputes. This perspective also gives a concrete outlook on research needs in biology. Regal *et al.* (1986) have given an impression of these basic research needs: "Much fundamental information is lacking about the life cycles of the vast majority of the specific forms in nature, about their evolutionary relationships and patterns of adaptation, about genetic exchange and stability in nature, about community organization and stability in nature, and about the linkages between community organization and the critical functions that microorganisms have in maintaining soil fertility, water chemistry, and atmospheric chemistry" (Regal *et al.* 1986: 3).

4.3.2 Relevant questions on receiving ecosystems?

Consider the following example of how an increased level of knowledge about biological complexity will help us recognize relevant research questions for GEO hazard identification. Bergelson (1994) has dedicated a study to the relevance of research questions to be included in a sufficient SRQ for GEO *fitness* identification. "In theory", Bergelson reflects, "fitness evaluations should predict the invasion success (or lack thereof)..." (Bergelson 1994: 249). This would imply that changes in invasiveness would be related to changes in fitness (operationalized by Bergelson as fecundity or seed production). An assumed SRQ for the purpose of invasiveness identification would therefore have to include questions about the relation between seed production and invasiveness as an important concern for consideration. Experimental design based upon such an assumed SRQ would have to focus specifically on *enhanced seed production* as relevant indicators of *enhanced invasiveness.*

The research done by Bergelson (1994) leads him to suggest an additional research question as relevant for the study of GEO invasiveness. His conclusion is that: "Changes in fecundity do not predict invasiveness" (Bergelson 1994: 249). As an alternative relevant research question to consider, Bergelson puts forward the relation between the GEO studied and the available *space* it encounters in its development: "I hypothesize that the absence of a detectable difference in invasiveness can be traced to the fact that space rather than seed production limits recruitment" (Bergelson 1994: 251). Thus, by introducing a new hypothesis as a basis for GEO hazard identification, Bergelson adds a new candidate concern to a sufficient SRQ for the purpose of GEO invasiveness identification. Current SRQs for detecting changes in invasiveness do not explicitly address the availability of space for a GEO: "The assumption underlying these trials is that changes in the performance of individuals will be reflected in changes in the invasiveness of species" (Bergelson 1994: 249). Where "space" may be limited in one ecosystem, it may be plentiful in another ecosystem – opening the possibility of inadvertent colonization.

QUESTION:

What is the *adaptive potential* of a transgenic organism?

Relevant because...	*Not relevant because...*
"In an *individualistic* approach adaptation of a GEO may depend on many relevant factors"	"In a *balance of nature* approach the fate of a GEO depends on available *ecological niche*"

QUESTION:

What is the effect of different *environments* on the expression of a transgene?

Relevant because...	*Not relevant because...*
"Different assessment environments or *microcosms* may trigger different GEO phenotypes"	"Only environment of intended release is relevant for assessment of transgene expression"

QUESTION:

What will be the available 'space' for the GEO in the target environment?

Relevant because...	*Not relevant because...*
"Available space rather than seed production limits GEO invasiveness"	"GEO invasiveness will depend on GEO *seed fecundity*"

FIGURE 4.2: Schematized representation of the scientific burden of proof for inclusion or exclusion of specific research questions as relevant or not relevant part of an SRQ for the hazard identification of a transgenic organism (no claim for *completeness* implied).

In the presentation of his results, Bergelson (1994) touches upon the *burden of proof for complexity reduction* in hazard identification: "My results highlight how difficult it can be to anticipate the invasiveness of particular plants or genotypes, and caution us against relying on simple "performance tests" as predictors of invasiveness" (Bergelson 1994: 251). This warning implies that the *scientific burden of proof* should be with those claiming that a field test based on a more simplified and less inclusive assumed SRQ provides relevant data for the purpose of GEO hazard identification. Bergelson (1994) has argued that research questions in relation to *available space* should be included as a relevant consideration in a sufficient SRQ for this purpose (cf. Lenski 1993: 202).

A fundamental controversy on *generic review* of GEO applications would deal with the *relevance* of specified research questions (see Figure 4.2). Both generic *and* case-by-case reviews must be based on a sufficient SRQ for a specified research purpose. The discussion should focus on the *(ir)relevance* of questions rather than on generic review exemptions.

— 4.4 Transgenic *micro*organisms —

An important special concern of GEO hazard identification is the understanding of genetically engineered *micro*organisms. Consider, for example, the following concern: "One of the hallmarks of general bacteriology has been the statement: 'Everything is everywhere, the environment selects'. What this means is that microorganisms have very minor problems getting around the world, in dispersing readily to distant environments. Therefore, it is not the presence or absence in a given environment which is critical, but the environmental conditions present in that habitat" (Brock 1985: 178). If this observation makes sense, then what does it imply for the relevant research questions for the purpose of GEM (genetically engineered microorganism) hazard identification?

One of the contested claims in the biosafety debate holds that it would be impossible for us to add something new to what has already been developed in nature: "What scientists create through genetic engineering is minuscule and ecologically insignificant compared to what occurs continually and randomly in nature" (Brill 1985: 118). Acceptance of this claim would affect biosafety regulation: "Unfortunately, from a regulatory point of view it is easy to think of possible harmful effects and ignore the presence of the same genes in organisms that already exist in nature" (Beringer and Bale 1988: 274).

There is no empirical test to assess the validity of this claim. Still, we can reconstruct what would be a sufficient SRQ which could lead to a justification of this claim that the impact of genetic engineering would be "miniscule and ecologically insignificant". Which research questions must be addressed to give this claim a scientific basis? Consider the following biosafety claim from *Bac-*

terial Genetics in Natural Environments (Fry and Day 1990). The core of this claim and its supporting assumptions can be found in different variations at many other places in the literature on biotechnology hazard identification (cf. Brill 1985, 1988; Jukes 1988; Gottesman 1985; Davis 1989). For an analysis and evaluation of the implied empirical claims, I reconstruct the underlying SRQ.

> "Because of the plasticity of natural bacterial populations and the importance of gene flow, which will almost certainly result in genetic modifications, it seems almost impossible that man could construct functional nucleotide sequences in genetically engineered bacteria that have not already been developed in nature. For these reasons our attempts at DNA technology must be pathetic compared with the process of natural selection in bacterial populations. Thus the chance that we could modify a bacterial genome in a way that would be harmful to the environment is vanishingly small. These arguments would suggest that we should not be too concerned about the release of genetically engineered bacteria into the environment" (Fry and Day 1990: 249).

Fry and Day (1990) are probably quite right to consider human efforts at recombinant DNA technology as more or less "pathetic" in comparison to what nature does. However, this does not justify a reassurance to the effect that we, "should not be too concerned about the release of genetically engineered bacteria into the environment". In my analysis, those contesting this claim essentially add relevant questions of concern to the sufficient SRQ for justifying this claim.

4.4.1 Already developed?

Which questions need to be addressed to claim that, "it seems almost impossible that man could construct functional nucleotide sequences in genetically engineered bacteria that have not already been developed in nature"? Consider Regal's analysis of this claim: "It is commonly argued that over hundreds of millions of years of evolution surely all combinations have been tried. (...) This idea is astonishingly simplistic and theoretically unsound. Take only *the first step* in a calculation: one human being alone, heterozygous at 6700 structural gene loci, can produce 10^{2017} different kinds of gametes. All the theoretical possibilities will never be realized even for the alleles of this restricted set of genes. The number of atoms in the known universe has been estimated at only 10^{70}" (Regal 1986: 120). As a particularly distressing example of this insight, consider the following observation: "The premisse that all possible gene combinations have already been tested in nature cannot be true (...). The sudden appearance of the virus that causes acquired immune deficiency syndrome [AIDS] should serve to convince us that nature occasionally does produce something with 'new' and unanticipated properties" (Sharples 1987: 1330). Perhaps the specific gene combinations constituting the virus have actually been around earlier somewhere. But the argument shows that the relationships between a microorganism and its host may be subject to change and cause very undesirable surprises.

Not including questions about those *developing relationships* in the conception of *'all possible bacteria have already developed'* as relevant part of a SRQ underlying this claim, seems to be a recipe for future disappointments. Questions about local interactions between (micro)organisms should be considered before a generalizing claim as the one made by Fry and Day (1990) can be justified.

The required insight to incorporate possibly relevant questions about these impacts conclusively in our considerations is still lacking: "It is not widely recognized that only 1-10% of the microbial populations of soil (and fewer from waters) have been isolated and cultured. Because these populations are unknown and uncharacterized it is difficult to predict their role in global cycles, the stability or invasibility of their genomes, their interactions with other organisms, and whether they are likely to be affected by introduced novel organisms" (Regal *et al.* 1986: 11).

4.4.2 Not harmful?

An implied claim in the argument by Fry and Day (1990) cited above is that the release of a GEM will not be harmful if its natural counterpart has already developed and has caused no harm: "Bacterial populations would not be in such harmony with the environment if specific combinations of bacterial genes represented a threat to the balance of nature" (Fry and Day 1990: 249). What research questions could be relevant to include in the SRQ that is a prerequisite to give this claim a scientific basis? To substantiate claims about hazards, we cannot get around a specification of particular impacts and agent(s) effecting them.

There is no mention, for example, of the *local numbers* and *density* of specific bacteria. What if applied biotechnology gives rise to *concentrations* of specific genetically engineered bacteria that would not occur in natural situations, thereby possibly overcoming a threshold of competition or reach a new adaptive peak? Depending on the specific GEM application, increased numbers of bacteria may be concentrated at one place, thereby enhancing survival chances compared to natural situations: "... what is different in the case of applying a microbial pesticide, for example, or a soil treatment of some kind, is the density at which these organisms suddenly occur in the environment" (Colwell 1985: 58). Thus it would not only be necessary to consider the genetically engineered organism by itself, but also the *numbers* in which it will be applied locally.

The competitive impact of GEMs applied in large numbers is, "... very different from what happens under normal evolutionary circumstances when a new genotype arises in a very, very low concentration and has a much harder uphill climb" (Colwell 1985: 58). Fry and Day (1990) do not address this concern in their argument and thus fail to include it in the SRQ from which to reason about this subject. Theirs is the burden of proof for showing that these questions need not be considered.

As an example of this concern, consider the following. One of the earliest field tests of genetically engineered microorganisms involved a genetically engineered *Pseudomonas syringae* increasing the *frost tolerance* of plants sprayed with it (Lindow 1985, 1990). The fact that this so-called "ice minus" GEM does also occur in a natural form, has led some to conclude that *therefore* its application does not pose a hazard: "... scientists are already beginning to agree that some kinds of experiments need less rigorous screening than others. These include experiments that, like Steven Lindow's so-called ice-minus bacteria, simply duplicate with recombinant DNA techniques an organism nature has already produced" (Tangley 1985: 472).

However, the isolated fact that the product of genetic engineering already exists in nature does not lead automatically to the conclusion that any planned introduction will be harmless. The possibility still exists that the release occurs in a place where it did not occur before or that it is released in numbers that enable the GEM to cross a specific threshold above which impact will be noticeable. For example, what if not only plants but also agricultural pests such as insects or weeds take advantage of the larger numbers of *ice minus* bacteria? This could lead to a situation where the cure is worse than the disease.

In the claim under study there is no specific information about what *type* of harm should be considered. It would be inconceivable to know the possible harm of an organism without considering its environment. The effect of a specific trait upon another organism depends upon the susceptibility of the affected organism. Perhaps a perfectly benign microorganism could replace other perfectly benign microorganisms and thereby cause changes in a microbial ecosystem that eventually lead up to unwanted effects. If this has happened in natural environments, then the chances are that we would not know about it: "... not all introduced microorganisms are pathogens, and the degree to which ecological modifications have been wrought by non-disease producing exotics will probably never be known" (Sharples 1983: 45).

4.4.3 No vacancies?

Other relevant research questions that need to be considered in GEM hazard identification pertain to the environment that a released GEM will find: "Even if all genetic combinations have been tried during evolution (and surely they have not) they could not have all been tried in all environments and ecological associations and across all the phyletic lineages of the last three billion years of life's history. Quite simply, not all genes were present in all places at all times" (Istock 1991: 125). In a useful SRQ for GEM hazard identification, this large number of possible options and their potential impact should not be overlooked.

Implicit in the claim by Fry and Day (1990) is the idea that harmful microorganisms will be eliminated in natural environments. How can we assess this? What questions must we include in a SRQ to evaluate this assumption? Fry and Day give the following impression of what they believe is going on at the level of microbial ecology: "Bacterial populations respond in hours and all the available bacterial niches in an environment are probably occupied in any one point of time" (Fry and Day 1990: 248). Thus, according to their view, the bacterial newcomers will find no *ecological vacancies* to live in.

One concern in the evaluation of this assumption is the implicit definition of the niche-concept, that was already criticized in Section 3.3.1. As Levins and Lewontin (1985), among others, have argued: an ecological niche is in fact an *interplay* between an organism and its environment. Our definition of a 'niche' will have marked implications for conclusions about the impact of ecological 'elimination' in natural environments. It seems difficult to evaluate the plausibility of the given claim without addressing this element of *ecosystem dynamics* and including it in the SRQ upon which we base our conclusions (see Figure 4.3).

Specifying these research questions will help us see the feeble scientific basis of biosafety claims on GEM release such as the following: "..., if it grows [a bacterial recombinant] more slowly than its competitors, by even an infinitesimal amount, the release of tons of the organism (whether deliberate or accidental) will have only a temporary and local effect" (Davis 1987: 1332). To evaluate this claim we must specify the empirical questions that need to be addressed to support it. One relevant question to ask in relation to this claim is *how temporal* and *how local* the effect would be precisely. Even a relatively short and confined ecological impact may disrupt an ecosystem on a larger scale and over a longer period of time.

QUESTION:	
What may be expected in relation to the *ecosystem dynamics* of transgenic microorganisms?	
Relevant because...	*Not relevant because...*
"Virtually nothing is known about microbial ecology"	"GEMs with *slower growth* than competitors will only have temporary and local effect"

QUESTION:	
Do GEMs which duplicate microorganisms already existing in nature pose possible hazard(s)?	
Relevant because...	*Not relevant because...*
"*Local density* of (transgenic) microorganism may overcome evolutionary thresholds"	"Everything is everywhere, the environment selects"

QUESTION:	
Could new GEMs pose new hazards?	
Relevant because...	*Not relevant because...*
"Not all ecological associations have been tried"	"Almost impossible to construct GEM which has not already been developed in nature"

FIGURE 4.3: Schematized representation of the scientific burden of proof for inclusion or exclusion of specific research questions as relevant or not relevant part of an SRQ for the hazard identification of a transgenic microorganism (no claim for *completeness* implied).

These and other questions must be addressed before the claims by Fry and Day could be considered 'scientific'. If Fry and Day (1990) would hold that these questions can be excluded from consideration and need not be included in a SRQ for the purpose of GEM hazard identification, then the methodological burden of proof for justifying this complexity reduction lies with them. Many research questions need to be addressed to give a scientific basis to such general claims as the one put forward by Fry and Day.

The challenge of finding these relevant research questions lies mostly before us, rather than behind us. In relation to the microbial ecology of *Bacillus thuringiensis*, for example, Hengeveld has expressed concern that: "Sadly, despite all our knowledge of its genetics, its physiology, and the way it interacts with its host (...), virtually nothing is known about its ecology" (Hengeveld 1994: 75).

— 4.5 Probability in an evolutionary context —

An interpretative concept that is a central concern in the context of GEO hazard identification is the "likelihood" or "probability" of some unwanted effect occurring. Understanding its methodological basis is an important prerequisite for understanding GEO hazards.

Consider a preliminary example. The safety concerns for applying GEOs have been distinguished as *biological containment* and *physical containment*. However, in practical circumstances, "contained use" can hardly be guaranteed. We may raise the question: what is the *likelihood* or *probability* that some GEO will escape from its physical containment, leading to an accidental release of the GEO and thereby to inadvertent exposure? Some applications may be less 'contained' than one might think. According to Stephenson and Warnes (1996), "... few commentators seem to realise that using naked DNA, genetically modified viruses or bacteria for gene therapy or vaccination will *inevitably* result in its release into the environment" (Stephenson and Warnes 1996: 12 – italics added; c.f. Canter Cremers and Groot 1991a, 1991b). In such a case, most of the concerns associated with a deliberate or intentional release would become relevant.

Claims about the (im)probability of accidental escapes must be validated against some interpretation of likelihood. *Exposure* is a condition for likelihood, but how well can we control 'exposure'? Wheale and McNally are pessimistic about the possibility of making *contained use* full proof and have dubbed the term "contained release" to express their concern about the imperfectness of physical containment: "We conceptualise contained use and deliberate release as being along a continuum of containment, ranging from high containment measures designed to prevent escapes, measures designed to minimise escape, measures which are simply good microbiological practice, through to controlled release and finally to open field release" (Wheale and McNally 1996: 188). This implies that concerns about *biological containment* should always be considered. On the other hand, field tests for the study of biological containment have been so restrictive that Bergmans (1995) has expressed his concern about their "field containment", making experimental results of dubious relevance.

Probability in an applied scientific context such as GEO hazard identification is much more than plain 'statistics'; it is about biological relationships. Whereas statistics is more concerned with the *quantitative* aspects of likelihood, probability is more concerned with its *qualitative* aspects. The *chance* that a likehood estimation is off target may be larger than the estimated likehood itself.

4.5.1 Independent chances?

In Section 2.5, we saw how the US *National Academy of Sciences* has concluded in a 1987 report that: "The possibility that minor genetic modifications with r-DNA techniques will inadvertently convert a nonpathogen to a pathogen is (...) quite remote" (NAS 1987: 15). To arrive at this conclusion, the NAS report urges us, "to recognize that virulent pathogens of humans, animals, and plants possess a large number of varied characteristics that in total constitute their pathogenic potential" (NAS 1987: 15). It goes on to list an impressive array of traits that are required for an organism to become pathogenic. On the basis of the striking discrepancy between making a "minor genetic modification" and the requirement of "a large number of varied characteristics" to arrive at a pathogenic potential, the inference is made that any "minor" action surely cannot lead to a major effect, such as pathogenicity.

In this line of reasoning, there are some interesting hidden assumptions (cf. van Dommelen 1996a). One such hidden assumption is that the modified organism to be assessed for a pathogenicity hazard potential does not already possess a substantial part of the necessary traits to become a pathogen before modification. The assumption that needs to be addressed here is that the array of required traits are supposed, in this argument, to be *independent* requirements. This is what makes it so "impressive" if they occur together. The probability of the development of inadvertent pathogenicity is the mathematical product, according to the NAS argument, of the chances of many different traits to develop out of a relatively minor modification. Of course, should the chances of the respective required traits somehow be *causally related* or *biologically linked*, then claims about the "remote" probability of a pathogenic potential to develop would become much less compelling. An important question to address then becomes: what exactly do we mean by the concept of 'probability' in relation to biosafety assessments?

Wimsatt (1984) has suggested to use the image of a, "bicycle lock with ten wheels of ten positions each", for the purpose of thinking about probability. In a normal situation, he explains, one would expect to need about half of the possible 10^{10} combinations before finding the combination which will work.

But what if the situation is not entirely 'normal'? Consider the clarification given by Wimsatt (1984) of the implied methodological challenge: "On the other hand, suppose that the lock is cheap or defective and one can tell individually for each wheel when it is in the right position. Then an average of 5 tries on each wheel, for a total of 50 tries, would be expected to find the right combination. The advantage that accrues from being able to break the problem down into subproblems, being able to find out parts of the combination, rather than having to solve the whole problem at once, is given by the ratio of the number of alternatives which must be expected. This is, in this case, $(5 \times 10^9)/(5 \times 10) = 10^8$" (Wimsatt 1984: 90).

This aspect of calculating probabilities may be relevant to the assessment of biosafety also. What if there is some evolutionary 'preference' for certain combinations in the 'lock' to arise, generated by biological dependencies. If, for some reason, the individual 'wheels' of nature's combinations are somehow not 'neutral' in their operation, then this is something to account for in claims about biosafety. If (part of) the wheels are causally related or otherwise have a preference for certain positions, then this will contribute to a higher probability for certain combinations to occur.

Alexander (1985) has promoted, in an oft-cited paper on biosafety assessment, a system to see: "[t]he probability of a deleterious effect [as] the mathematical product of the probability of release, survival, multiplication, dissemination, and actual harm", and concludes: "Hence, the risk of genetic engineering is probably small" (Alexander 1985: 64). However, what if these probabilities are not biologically independent? Then, in parallel with the defective bicycle lock, a different likehood will be the result. Regal (1988), for example, has conceptualized the possible evolutionary effect of a genetic insertion as follows: "The addition may set in motion a 'hunt' for a new equilibrium. So the genetically engineered organism having a new adaptive trait with a positive selective value may have inherent properties of unpredictability, associated with this 'hunt'" (Regal 1988: s38). The possible relationship between seemingly independent chances then becomes a relevant question to address. Those leaving these questions out of consideration as being not relevant thereby take on the burden of proof for the methodological adequacy of this complexity reduction.

4.5.2 Early warning?

Thompson (1987) has drawn the consequence of the line of reasoning that Alexander (1985) suggests: "Any harmful effects that *did* occur from such an action would have to be classified as *so* unexpected that they are, in effect, freak accidents, the sort of outcome for which no one can be held accountable" (Thompson 1987: 322). Perhaps we would be well-advised to think about the possible hazards of the environmental release and agricultural use of genetically engineered organisms, in terms of possible "normal accidents" rather than in terms of possible "freak accidents". Hazards are often seen as 'hypothetical' or 'conjectural' in safety assessment more generally. Until some *accident* occurs, then suddenly the safety becomes hypothetical. Perhaps our interpretation of this must be that in those cases the projected 'freak accident' turns out to be a 'normal accident' (Perrow 1984).

Those who claim that there is reason to think that the likelihood of a specific hazard to occur will be close to zero, will have to specify their concept of *probability* in this context. A complication in quantifying biological possibilities is the fact that biological entities may change and adapt through processes of biological *evolution.* As long as we cannot specify the relevant questions for GEO hazard identification, we cannot say much about the probability of the occurrence of an unwanted effect and must be prepared to consider its possibility.

For example, the occurrence of *weediness* may be studied in terms of its probability of occurrence (cf. Keeler 1989, Fitter *et al.* 1990). According to Williamson *et al.* (1990) the probability for weediness may be much higher than Keeler (1989) has argued: "With so many weeds related to crops, a reasonable guess is that the probability of changing a randomly chosen crop into a weed by changes in distribution or cultivation practice, or by natural selection, at some time and in some place, is perhaps 10^{-1} to 10^{-2}, not 10^{-10} as suggested by Keeler (1989) – a difference of eight or nine orders of magnitude" (Williamson *et al.* 1990: 418).

Consider a contested probability claim in the context of GEO hazard identification to which the possibility of overlooking an effect is particularly relevant. According to some there exists an "early warning principle", which would work because: "Statistical laws imply that pests which cause great harm could not be inadvertently produced by genetic engineering from innocuous organisms without being preceded by an *early warning* consisting of the appearance of some weakly harmful constructs" (Szybalski 1985a: 115 – italics added, cf. Szybalski 1985b). Davis (1989) similarly argues that, "..., built into any realistic scenario for adventitious harmful spread of a recombinant is an early warning system before serious problems would arise. Since the enormous experience with nonpathogenic recombinants in the laboratory has not given us any trace of a warning, it seems reasonable to conclude that the chance of serious harm is not only small: it is negligible" (Davis 1989: 868).

QUESTION:

Are the separate factors determining a possible GEO hazard *biologically independent*?

Relevant because...	Not relevant because...
"If determining factors of a hazard are not independent this may increase likelihood of occurrence"	"Probability of hazard is *mathematical product* of individual chances for determining factors"

QUESTION:

What is the probability of GEOs 'escaping' from *physical containment*?

Relevant because...	Not relevant because...
"Physical containment is on a *continuum of containment* and will not be absolute"	"*Contained use* will not require biological containment"

QUESTION:

What detection methods can predict the occurrence of a possible GEO hazard?

Relevant because...	Not relevant because...
"Probability of occurrence will depend on specific release conditions"	"*Early warning* by the appearance of weakly harmful constructs"

FIGURE 4.4: Schematized representation of a biosafety controversy over *probability*. Vague terms such as "early warning" and "probably small" add to artificial disputes. Reconstruction in terms of relevant research questions brings back the discussion to the fundamental controversy over the choice of a sufficient SRQ for the hazard identification of a transgenic organism (no claim for *completeness* implied).

These claims completely overlook the importance of experimental design in the detection of any kind of early warning. They *assume* the 'visibility' of some preceding or small-scale effect without clarifying the methodological basis of this claim. Another matter is whether an early warning would be *early enough* to take adequate measures. What would have been the "early warning" for such an "explosion in slow motion" (Tenenbaum 1996: 34) as that of the spread of

Australian rabbits or of Kudzu vine or of waterhyacinths, etc.? These experiences have shown that even 'early' warnings may be too late. What are the relevant research questions for detecting an "early warning"? What are the circumstances deciding whether or not an "early warning" will occur?

To replace *artificial* controversy over the probability of GEO hazards with *fundamental* controversy, we must be prepared to argue for the *(ir)relevance* of specific research questions in relation to the purpose of GEO hazard identification (see Figure 4.4). Claims about "early warnings" or "negligibility of chances" of GEO hazards to occur can only be evaluated against the background of an appropriate interpretation of *evolutionary probability* and the associated relevant research questions.

CHAPTER 5

The 'Familiarity' Criterion in Biosafety Regulation

— 5.1 Enough information? —

In the previous chapters, it was argued that biosafety controversies can be analysed and evaluated as disagreements about the *relevant research questions* to be raised for the hazard identification of genetically engineered organisms (GEOs). Where scientists disagree, politicians may be expected to look for pactical ways of coping with the consequential lack of scientific certainty. In relation to the biosafety assessment of planned environmental release of GEOs, policy makers have introduced the notion of 'familiarity' for this purpose. In this chapter, a proper interpretation of this notion is discussed and a practical approach is presented to give 'familiarity' a scientific basis without which it should not be used in a policy context.

Depending on the accepted definition of 'familiarity', different scientific procedures for biosafety assessment will be considered satisfactory. Biosafety tests such as specified by the 'Flowcharts' as developed by the ABRAC (1995a, 1995b), reflect a view on familiarity also. To develop such 'Flowcharts' for biosafety testing on the basis of up-to-date science, a perspective on familiarity is needed that allows for adjusting it to developing insights. This important element of *adaptative flexibility* seems to be missing from most uses of this convenient term. To compensate this omission, below a *procedural scientific approach* to the use of the familiarity criterion in biosafety policy is presented. Agreement over the relevant questions for 'familiarity' may give scientific legitimation to political decisions. Specification of these questions will therefore be in the interest of scientists as well as policy-makers and politicians.

The notion of 'familiarity' has been propagated as a new policy tool to bridge the gap between scientific expertise and regulatory practice in the context of biosafety assessment of genetic engineering (cf. UNCED 1992; ABRAC 1995; Doyle and Persley 1996; UNEP 1996). As was argued in Chapter 1, policy makers are in need of such bridges. The *Organization for Economic Cooperation and Development* (OECD), which has been concerned with managing the risks of biotechnology for many years now, has chosen the *familiarity criterion* as an organizing concept in its policy on research and regulation for biosafety assessment. The OECD (1995) states: "The concept of familiarity is a major factor in all phases of the evaluation, since it is used to identify potential adverse effects (*i.e.*, hazard identification), to determine the level of risk associated with these adverse effects, and to adopt risk management strategies" (OECD 1995: 18). The present analysis can be a useful tool to assess the scientific merits of the concept of familiarity. The criterion is designed to amalgamate normative and scientific aspects of the biosafety issue. Understanding the *scientific meaning* of 'familiarity' in this context is a prerequisite for appreciating its normative role in regulatory decision-making.

The earliest proposal to use the familiarity criterion as a descriptive strategy for dealing with biosafety issues was made by the US *National Academy of Sciences* (NAS 1989). It recommended breaking down the larger biosafety question into smaller parts and to begin with answering the question: "Are we *familiar* with the properties of the organism and the environment into which it may be introduced?" (NAS 1989). In cases where this question can be answered in the positive, it is implied that considerable ground has been gained in the larger risk assessment question.

The NAS (1989) continues: "When the familiarity standard for a plant or microorganism has been satisfied such that reasonable assurance exists that the organism and the other conditions of an introduction are essentially similar to known introductions, and when these have proven to present negligible risk, the introduction is assumed to be suitable for field testing according to established practice" (NAS 1989: 5). Given the uncertainties that surround biosafety assessments, this approach presents a convenient option for the regulatory authorities. The NAS (1989) phrases the advantage of dealing in this way with the involved scientific uncertainty as follows: "The familiarity criterion is central to the suggested framework of evaluation. Its use permits decision-makers to draw on past experience with the introduction of plants and microorganisms into the environment, and it provides future flexibility" (NAS 1989: 5).

To evaluate whether or not the 'familiarity criterion' is any good in this context, sufficient understanding of how it can be defined and applied is needed. What are the prerequisites to evaluate or assess 'familiarity'? As yet, the notion of familiarity has not been spelled out in sufficiently scientific terms to serve the purposes envisaged by NAS, OECD, and other institutes. Considering the terms in which 'familiarity' has found its way into the regulatory landscape of biosafety assessment, the conclusion must be that it hinges on vague descriptions such as "essentially similar" and "reasonable assurance" (NAS 1989). The OECD (1995) has provided equally vague qualifications: "Familiarity is not synonymous with safety; rather, it means having *enough information* to be able to judge the safety of the introduction or to indicate ways of handling the risks" (OECD 1995: 12 – italics added). What would be a more scientific definition of 'familiarity' in this context?

5.1.1 NEED-TO-KNOW VERSUS NICE-TO-KNOW

As is stated by the OECD, "familiarity does not imply safety, it implies availability of information" (OECD 1995: 18). A basic prerequisite for assessing the "availability of information" is to know *what* information is wanted. As we have seen in the previous chapters, there is controversy about what constitutes relevant information or relevant questions for biotechnology hazard identification. A distinction has been proposed between what we "need to know" and what would be "nice to know" for the purpose of GEO hazard identification (CCRO 1995: 4). Accordingly, the underlying challenge for a scientific application of the familiarity criterion has been characterized as: "More consensus about 'need-to-know' issues versus 'nice-to-know' issues would seem to deserve the utmost priority" (Metz and Nap 1997: 44). Biosafety controversies are about the contested relevance of research questions for the purpose of hazard identification. The same challenge must be faced to arrive at a scientific use of the 'familiarity' criterion.

The OECD has also promoted the view that biotechnology risk assessment should be based on the "best available knowledge" (Zannoni 1995: 36). To live up to that scientific standard in relation to the 'familiarity criterion', we cannot get around answering the question: *what knowledge is required (i.e., what would be "enough information" or "reasonable assurance") to decide that different applications of biotechnology are "essentially similar" in this context?* Answering this question is a prerequisite for making the notion of 'familiarity' operational for its scientific purpose in biosafety assessment. The everyday appeal of the notion of 'familiarity' could tempt its users to fall into the trap of having a simplified understanding of the biological complexity involved. Methodological precaution must be taken to make sure that users of the 'familiarity criterion' do not in effect resemble the well-known drunk who is looking for his lost keys under the lamp post, "because there is more light there". Uncritical interpretations of 'familiarity' could easily lead to a neglect of possibly relevant hazard concerns.

It should be noted that "enough information" has a scientific as well as a political or administrative connotation. In one interpretation it can mean that 'enough' information is available to *support a biosafety claim.* In another interpretation it can mean that the available knowledge is judged a sufficient basis to *go ahead and take a societal risk.* The latter is a daily challenge for public administrators and therefore nothing new. However, it should not be legitimated with a false sense of scientific certainty based on a misguided use of the first interpretation of 'enough'.

— 5.2 Interpreting familiarity —

Thinking about science as a special kind of detective work and about biosafety assessors as applying a scientific searching device to do their work, it may be asked how useful a tool the concept of 'familiarity' can be for GEO hazard identification. Of course, one can never claim to have familiarity with the so-called 'unknown unknowns' of applied biotechnology. The basic challenge of assessing familiarity in relation to biosafety issues can be phrased as: what are the research prerequisites to decide whether or not to include a specific 'known unknown' in hazard concerns? A scientific interpretation of the notion of familiarity must address the relevance of including or excluding specific hazard considerations. It requires that reasoned support is provided for claiming that some recognized unknown can be interpreted and treated as sufficiently similar to what we already know from another context.

Francis Bacon (1620), who is often referred to as one of the founding fathers of modern experimental science, has warned us against the human tendency to interpret the world as more familiar than is warranted:

> "The human understanding is of its own nature prone to suppose the existence of more regularity in the world than it finds. And though there be many things in nature which are singular and unmatched, yet it devises further parallels and conjugates and relatives which do not exist" (Bacon 1620: 51).

From a scientific perspective, the notion of 'familiarity' may be too general a concept to serve the purpose of reducing uncertainties and rendering visible what was previously unseen in biotechnology risk assessment. By claiming familiarity where it turns out to be not warranted, we are handing down our present scientific uncertainties and ignorances (and their potential consequences) to future generations.

'Similarity' and 'comparison' are important notions in applications of the familiarity criterion. From a practical perspective, this may be reason for enthusiasm: "Familiarity as a tool can be strongly recommended, since it relates the assessment to a comparison with known practice (...) For instance, the intended release of modified rhizobia should be weighed against the history of about one century of safe releases of rhizobia to field soils ..." (van Elsas 1995:

125). However, more skeptical interpreters can also be found (cf. Wittgenstein 1965, 1971). Williamson (1994) is one of the participants in the biosafety debate who has expressed doubts about the scientific usefulness of this criterion: "Familiarity is unlikely to be an effective defence against new ecological effects (...) Our familiarity with these species as useful agricultural and horticultural plants may be irrelevant and misleading" (Williamson 1994: 75). A prerequisite for any scientific application of the 'familiarity criterion' is a proper definition and interpretation.

One important defining element of 'familiarity' is *comparison* (see also Section 3.4). Assuming that enough knowledge is available about conventional organisms and their applications, the basic idea of the familiarity criterion is that transgenic organisms have a hazard potential that is *comparable* to nontransgenics. Apart from possible doubt about the prior assumption of presently available sufficient knowledge about previous agricultural applications, the strategy of comparison implies facing the task to specify which aspects of a conventional release and a transgenic release are *relevant* for comparison (given the objective of hazard identification). It requires some *perspective of comparison* as a basis to design reliable safety experiments. Since safety cannot be expressed in absolute terms, a GEO can only be judged "as safe as" a familiar organism (Käppeli and Auberson 1997: 346). But then it must be specified what is considered a relevant comparison. The ideal perspective of comparison would be a specification of all the relevant empirical questions that need to be addressed to claim familiarity with the possible hazards of a GEO. To apply the familiarity criterion in a scientific way one cannot get around producing a *set of relevant questions* (SRQ) underlying the comparison between the use of conventional and transgenic organisms.

Most deliberate releases will probably not lead to trouble; the challenge of GEO hazard identification is finding ways to detect the minority that will. This implies that detection methods must be subtle enough to recognize this small group of cases. The notion of familiarity does not provide the required subtlety by itself. From a scientific perspective it is a *blanket-like* notion: covering many things without much distinction. One aspect of familiarity that should be taken into account for scientific use is that 'familiarity' is always relevant to some purpose of research. For example, one may be *familiar enough* with a clock to adjust its time, and at the same time *not familiar enough* to repair it. Maybe, in the context of GEO hazard identification, the first *level* of familiarity is to know which questions must be addressed. This is a much more modest research goal than having all the answers (see Figure 5.1).

Familiarity comes with a methodological burden of proof. It must be assumed that one is 'unfamiliar' with some application until proven 'familiar'.

The person claiming 'familiarity' has the burden of proof to justify the claim that the relevant questions for the purpose of hazard identification have been sufficiently addressed. This is a prerequisite for a scientific interpetation of the familiarity criterion. To specify familiarity, a way must be found to specify what are considered to be relevant questions. Making an *inventory* of the relevant questions is important even apart from possible answers. One may not know the answer to a question and still know that it needs to be addressed.

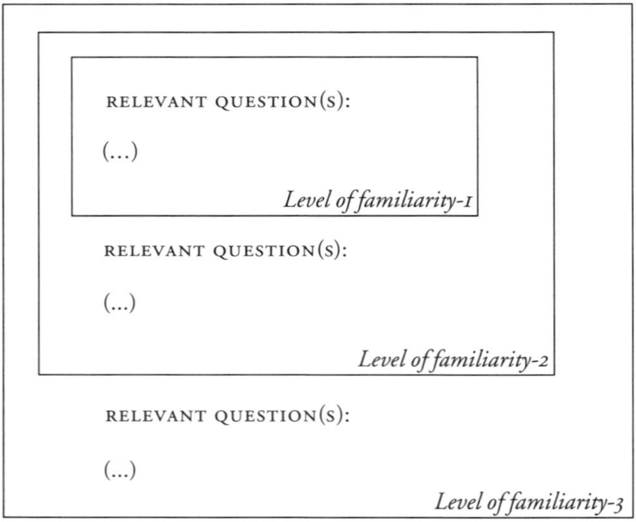

FIGURE 5.1 Schematized representation of a scientific interpretation of the familiarity criterion. Different levels of familiarity may be reached or required in relation to a specified research purpose. To allow a scientific discussion, familiarity must be interpreted in terms of the relevant research questions that could produce "enough information" for the purpose of GEO hazard identification.

5.2.1 AN INTERNATIONAL RESEARCH PROJECT?

The familiarity criterion is not so much meant as a tool to calculate risks, but rather as a tool to *recognize* possible hazards. If the familiarity criterion is to serve as a guide in the process of hazard identification, then it must help us to ask relevant questions. Given some level of exposure (for example, deliberate release of genetically engineered organisms), we are interested in recognizing the possible hazards in order to foresee possible risks. For this purpose, we need to produce an interpretation of 'familiarity' that can work as a *searching device.* The difficulty of achieving this task may be illustrated by the following observation: "Considering the current state-of-the-art and paucity of data on detection, enumeration, survival, growth and transfer of genetic information (both intra- and interspecifically) by genetically engineered microorganisms (GEMS) in natural habitats, the detection, measurement and evaluation of potential effects of an introduced GEM on ecological processes is like trying to find 'a needle in a haystack'" (Stotzky 1990: 147).

How much familiarity can now be claimed in relation to this "haystack" of biological complexity? Or, in other words, what would be *enough information* or *reasonable assurance* to claim that one 'haystack' is *essentially similar* to another? Addressing this question is complex and important enough to be made the focus of an *international research effort* of experts around the world for the purpose of supporting biosafety assessment (cf. Regal 1995a, 1995b). Familiarity may be operationalized methodologically as a measure for the need (or absence thereof) to do specific biosafety experiments (address specific empirical questions). Thus, one must specify which are the relevant research questions to address before one can claim familiarity.

— 5.3 Familiarity with horizontal gene transfer —

As an illustration of the methodological difficulties encountered in a scientific interpretation of the familiarity criterion, consider the following. The possible impact of *horizontal* or *lateral gene transfer* is an important concern for GEO hazard identification: "Serious concerns must remain (...) over the potential for genetic transfer in the environment. Mechanisms exist whereby any genetic trait can, in theory, be transferred between any two prokaryotic organisms and between prokaryotes and several eukaryotic species. It is essential therefore that all uses of any genetically modified organisms, no matter how apparently benign, and which involves its use outside the laboratory, must be carefully controlled and monitored. History may prove this attitude to be erring on the side of caution, but the rapid spread of antibiotic-resistant bacteria graphically demonstrates that a less cautious approach during the infancy of this technology could have disastrous consequences" (Stephenson and Warnes 1996: 13).

Concern about possible hazardous effects of horizontal gene transfer is especially interesting in relation to the use of the *familiarity criterion* because this concern has a somewhat ironic history. Until the beginning of the 1990s there was little concern about lateral transfer because it was thought to be virtually absent in natural situations: "To date, gene transfer from transgenic plants to microorganisms has never been detected" (CEC 1993: 31). However, new research led scientists to, "... hypothesize that in nature many species, if not all, have access to one large gene pool (...) Therefore, horizontal gene transfer may have substantial impact on the evolution of species" (Beijersbergen 1993: 101; cf. Beijersbergen *et al.* 1992). When it became clear that horizontal gene transfer is in fact a widespread phenomenon in nature, some took this as a *new* reason to downplay concern: 'If it is so widespread than it is a natural phenomenon anyway' (cf. Potthast 1996). It shows how discussions on biosafety can only be clarified by specifying our 'familiarity' in terms of the relevant research questions: "Genes travel between independent bacteria more often than once was assumed. Study of that process can help limit the risks of releasing genetically engineered microbes into the environment (...) A key to the safe release of the microbes (...) is to identify the conditions that will encourage or deter specific bacteria from transferring their genes to other organisms..." (Miller 1998: 47).

Käppeli and Auberson (1997) give an example of how the "familiarity principle" might be applied in practice. They consider the possible hazard of the inserted, "pest-resistant gene (Bt toxin gene) and herbicide-resistant gene (marker)" to 'escape' by *horizontal gene transfer* from transgenic cotton plants. In their view, application of the familiarity criterion helps us see that the respective uses of transgenic and nontransgenic organisms in this case pose a, "similar hazard potential" because the genetic information introduced into the cotton, "has always existed in the environment" (Käppeli and Auberson 1997: 347). To specify their argument, I cite a somewhat longer section: "The transformed and mutated soil microorganisms that may harbor the unwanted *Bacillus thuringiensis*-toxin and herbicide-resistance traits obtain these genes by gene transfer from naturally-occurring *B. thuringiensis* or spontaneous mutations under selective pressure introduced by the use of herbicide in the environment. Therefore, the hazard and damage from the emergence of transformed, mutated soil microorganisms retaining unwanted traits could result from pathways other than GEO releases; the trait[s] potentially transferable from the transgenic cotton plant to the environment do not represent novel genetic information for the surroundings, nor a measurable aggravation of an existing, tolerated background state" (Käppeli and Auberson 1997: 347).

What does the familiarity criterion address in this application? In other words: which are the empirical research questions considered relevant by Käppeli and Auberson (1997) for the purpose of transgenic cotton hazard identification? In my view, they step into the pitfall of claiming more familiarity than is warranted. By failing to specify possibly relevant empirical questions (or their assumed SRQ), they fail to identify possible hazard potentials. Below, I present some of the empirical questions Käppeli and Auberson (1997) omit from their concern, although they may be relevant for the purpose of their research.

Käppeli and Auberson (1997) state that the genetic information introduced into the cotton, "has always existed in the environment". This knowledge is certainly a relevant aspect of familiarity, but is it sufficiently specific? In my view it is not. One relevant empirical question to address in addition is: has the introduced genetic information "always existed" in the *same concentrations* and in the *same places* as will be the result of the environmental release of transgenic cotton? Higher concentrations of occurrence of a transgene may give rise to different ecological impacts. Since agricultural applications will typically result in large-scale and frequent introductions, the use of transgenic cotton and the resulting possibility of horizontal gene transfer may lead to the crossing of 'evolutionary thresholds' that were not crossed before (cf. Brandle *et al.* 1995). The effect of this could be that the use of transgenic cotton would *not* pose a "similar hazard potential" to the natural situation. Failing to address the empirical questions of *local density of genes* and *environmental distribution of genes* in relation to horizontal gene transfer may thus lead to overlooking a potential hazard.

Another empirical question that Käppeli and Auberson (1997) fail to address in their assessment of our familiarity with the possible effects of horizontal gene transfer resulting from the deliberate release of transgenic cotton, is the relative *genetic stability* of a transgene in the GEO. As Maessen (1997) has observed: "It may be assumed that transgenic plants, genetically, behave differently from their parents due to changes in the genome caused by the introduction of the transgene" (Maessen 1997: 4). According to Maessen (1997), different concerns are relevant to the assessment of 'familiarity' with this aspect of GEO hazard identification: "These factors include characteristics of the host genome, the transgene itself, the mechanism of introduction and the site of integration. An inventory of these factors should allow determination of whether the genome of a transgenic plant is likely to be stable" (Maessen 1997: 5). Changes in genetic stability may result in changes in the rate of horizontal gene transfer, and thus in changes in the hazard potential of a GEO. Since Käppeli and Auberson (1997) fail to address the relevance of the empirical questions articulated by Maessen, they cannot claim familiarity with the use of transgenic cotton *in this respect.*

The supposed familiarity with the hazard potential of horizontal gene transfer from transgenic cotton, as claimed by Käppeli and Auberson, should be qualified *against the background of the empirical questions they addressed* and failed to address. It seems to me that the burden of proof rests with Käppeli and Auberson to argue that the questions raised by Maessen (1997) are *not* relevant in this context. Familiarity is never absolute. Its measure in the context of hazard identification is: do we know which empirical questions are relevant to address in relation to the planned introduction of a GEO?

5.3.1 Specifying research questions

Apart from the *mechanism* of gene transfer, we may also be concerned with the *effect* of the transferred GEO traits. In an attempt to specify the relevant questions for the purpose of identifying the hazard of GEO *invasive weediness*, Williamson (1993, 1994) has addressed the questions: which traits of a GEO will determine, "whether the GMO [= GEO] will invade?" (Williamson 1994: 76) and "[i]s it possible to predict, from their characters, which invaders will become pests?" (Williamson 1993: 221). In the present analysis, this corresponds to asking: which biological traits are relevant to consider in the identification of GEO weediness? An assumed SRQ must provide us clues about what specific GEO traits to look for in our search for possible hazards. If, for example, we should know that a specific GEO under study can be *pathogenic* in some circumstances, this is obviously a good reason to give special consideration to this trait in any further hazard identification.

The way Williamson (1993) has operationalized his question whether invasive weeds, "have distinctive constellations of characters" (Williamson 1993: 221), is by taking twelve of the fourteen traits that were originally distinguished by the weed scientist Baker as defining the "Ideal Weed", and to see if any of these characters would specifically *predict* invasive weediness. This can be understood as an attempt to seek 'familiarity' with GEO weediness. Williamson's findings are not very promising in terms of specifying a sufficient SRQ for the purpose of GEO weediness identification. He concludes that: "Baker characters are not predictive. It is not necessary to add Baker characters to make a crop into a weed. (...) Adding or subtracting Baker characters leads to no useful prediction about weediness" (Williamson 1993: 222). Given this difficulty to be specific about a useful SRQ for this research purpose, Williamson (1993) suggests to take refuge to an empirical 'rule of thumb' for hazard identification: "... roughly 10% of invaders establish, become fully naturalized, and again roughly 10% of those become pests" (Williamson 1993: 219).

Although this so-called "ten-ten rule" may give biosafety assessors a general handle to regulate the environmental release of GEOs, they will not be particularly satisfied with this 'rule'. Biosafety regulators will want to know *which* ten percent of invaders will establish and *which* one percent will eventually become pests. To identify these categories we must depend upon further sophistication of our ideas about what constitutes a sufficient SRQ for the purpose of identifying GEO weediness (cf. van Lenteren 1992).

Williamson (1993) has argued that the famous list of "Baker characters" cannot by itself be the bleuprint for this SRQ. The suggested alternative, the assumed "ten-ten rule", is not uncontroversial either. According to Crawley: "The so-called 10-10 rule (...) is wrong for plants. Far fewer than 1 in 10 of the plants brought to Britain has become established and of those that become established, less than 1 in 10 become pests" (Crawley 1994: 44). Others have claimed about the percentages that, "[i]t is estimated that 5% of unintentionally introduced organisms establish, and that 7% of these become pest species" (van Lenteren 1992: 59).

What does this imply in relation to the methodological *burden of proof* for an assumed SRQ to identify the hazard of weedy invasiveness? Since Williamson (1993) has shown that it is not possible (as yet) to indicate which characters are especially relevant to consider, this implies that it is not possible on the basis of his results (as yet) to decide which of these characters can be *excluded* as not relevant from a SRQ for the purpose of weediness hazard identification (given their possible relation to weediness as established by Baker). Anyone claiming to have enough familiarity with GEO weediness to *exclude* from consideration and thus from an assumed SRQ any of the characters considered by Williamson (1993) thereby takes on the burden of proof for justifying this claim (c.f. van Lenteren 1992).

— 5.4 Familiarity with relevant research questions —

The concept of 'familiarity' can only be operationalized and used as a tool for biosafety policy by making explicit the SRQ for which 'familiarity' is claimed or sought. Thus, the first and foremost challenge for biosafety assessors is to develop a procedure that enables us to define our 'familiarity' with a GEO release in terms of the relevant research questions for the purpose of hazard identification. This should be a *dynamic* procedure in the sense that it should allow an ongoing process of re-definition in response to scientific developments and it should be a *practical* procedure in the sense that it should allow us to specify concerns without loosing ourselves in endless controversy. Such a procedure could follow the methodological lines of analysis and evalution developed in Chapter 7.

An important aspect of the *dynamic* character of 'familiarity' is the fact that modern biotechnology or genetic engineering is in many respects ahead of the developments in the biological sciences when it comes to biosafety assessment.

Thus, research questions that determine 'familiarity' will accumulate or be dropped in relation to the developments in the applied context of biotechnology. This dynamic character of 'familiarity' makes it the more important for policy-makers to focus on the *procedural* conditions or requirements to ensure an adaptive approach to new scientific knowledge about GEO hazard identification.

Apart from the dynamic aspect of the development of our knowledge, we must also be aware of the fact that 'familiarity' with one GEO release does not imply that the same SRQ will be useful for the hazard identification of another GEO release. In other words, we cannot expect to master some sort of *generic* familiarity or a *complete* SRQ. The level of insight will have to be specified for different contexts and applications. This complication may be illustrated by the following observation in relation to GEO biosafety assessment: "... environmental impacts estimated over one range of conditions may be of little use in predicting behaviour over a different range of conditions ..." (Crawley 1994: 36).

The practical analytical tool of an appropriate SRQ in view of a specific research purpose allows a detailed evaluation of *individual* relevant questions, thereby giving the procedure of defining 'familiarity' a *dynamic* and *modular* character. A focus on the relevant questions as the basis of familiarity claims may be a practical way to accommodate the demands of an adaptive and dynamic system of biosafety assessment. Maybe, in the future, we will learn to 'close in' the most essential or sufficient SRQs for the purpose of specific GEO hazard identifications. For the time being, a prudent approach requires us to adapt our 'window of concern' to the limited knowledge about relevant questions that is available now.

It is telling that the OECD itself does not put the familiarity criterion to any practical use in its own summary of case reviews (OECD 1995: 28-51). In the end, we must express our knowledge and lack thereof in scientific terms and therefore we need a *scientific interpretation* of 'familiarity'. As an intermediary between science and regulation, the notion of familiarity can certainly play a role.

Using it as a policy tool, we should be aware that familiarity can cut two ways. The burden of proof for claiming familiarity is asymmetrical. Claiming familiarity with the *un*safety of an intended GEO application can be quite convincing in some cases. For some applications we have sufficient understanding and experience to claim familiarity in relation to its possible negative environmental impact. The reason for this is that addressing *one* relevant question can be decisive for establishing a potential hazard; in contrast, *one* relevant question will never be a sufficient basis for establishing the safety of a GEO release. We are sufficiently familiar, for example, with known pathogens to justify the conclusion that there is no need for further assessment, because there is an obvious possibility of harmful effects. Claiming familiarity towards sufficiently assessed safety, however, puts rather different demands upon the required SRQ.

As a critical instrument, the notion of familiarity can be quite useful. But then one must try hard to include proper science and be prepared to acknowledge that, from a scientific point of view, full familiarity is a distant goal in most cases. An important element of hazard identification is deciding what to exempt from consideration. Biosafety experiments should be designed to uncover relevant empirical questions to address in biosafety assessments. This will help us gain understanding as opposed to being satisfied with collecting more and more data of dubious relevance.

As a constructive practical attempt to give the familiarity criterion a scientific basis, consider again the *flowcharts of relevant empirical questions* developed by the USDA *Agricultural Biotechnology Research Advisory Committee* (ABRAC 1995a, 1995b; for more detail, see Section 2.1). An important element of a scientific basis for the familiarity criterion is the recognition that familiarity comes in degrees, as the following may illustrate: "When a modified organism is first studied in confined experimental systems, familiarity with its overall phenotype would be expected to be quite low. After substantial phenotypic testing, the degree of familiarity could increase to the point where it becomes possible to give a clear affirmative or negative answer to the question about phenotypic changes" (ABRAC 1995a: 25).

In this approach, the relevance of questions is made an explicit concern of GEO hazard identification. For example, to claim a higher level of 'familiarity' with possible phenotypic changes of a particular GEO: "It is imperative that experiments involve proper measurements for these phenotypic changes and that inter-trait correlations and genotype-environment interactions be considered" (ABRAC 1995: 25). Thus, familiarity represents a developing level of knowledge which can always be contested. The SRQ approach developed and demonstrated in this study can serve as an aid for organizing the debate about a scientific definition of 'familiarity' with GEO releases in terms of sufficient SRQs.

The basic way to give a scientific basis to familiarity in this approach is to ask: do we know which questions should be included in a 'flowchart' or a SRQ for a specific purpose? An important advantage of this approach of explicitly listing relevant empirical questions for specific research purposes in 'flowcharts' is that the claimed level of familiarity will always be a reflection of the *sophistication of these flowcharts.* More detailed and sophisticated questions will potentially result in more qualified familiarity. More superficial and general questions, on the other hand, will result in a less qualified familiarity. This approach helps us to avoid application of the familiarity criterion as an inarticulate *blanket-like* qualification, which is useless if not dangerous in the context of GEO hazard identification.

5.4.1 Controversies as a source of relevant questions

How can biosafety controversies be interpreted in a constructive vein? In a study on, *The political influence of global* NGOs, Arts (1998) comes to a rather gloomy conclusion about the impact of critical parties on the outcome of international discussions about biosafety policy: "Several NGOs played a very activist role on biosafety, but so did several developing countries. Despite pressure from these two sides, the first session of the Conference of the Parties (COP 1) decided not to start immediate negotiations on a protocol, due to opposition from countries which either rejected any instrument (USA) or favoured voluntary and technical guidelines under the UNEP or the Commission for Sustainable Development (United Kingdom, Netherlands). Instead, COP 1 established a working group to consider the need for a protocol. The NGOs were disappointed about that. In addition, some of their other proposals were not taken up either (moratorium on GMOs). Given this outcome, NGOs did not exert any political influence on the biosafety topic in the period 1992-1994, although their interventions on the dangers of GMOs and the need for strong international biosafety regulations definitely fuelled the debate (process influence)" (Arts 1998: 220).

This account may underestimate the impact of interruptions made by NGO representatives in international discussions, but it draws attention to an important problem: how can we benefit from the critical efforts of NGOs worldwide? One open and practical road that I see is to use the work of NGOs in the context of GEO hazard identification as useful *resources of possibly relevant research questions.* By contributing to a scientific interpretation of the 'familiarity' criterion, the NGOs may exert a real influence on the development of biosafety policy.

It has often happened in the history of environmental impact assessment that the 'competent authorities' have had to acknowledge *ex post facto* that critical concerns expressed by NGOs turned out to be warranted. At present, we are seldom sufficiently equipped to claim scientific familiarity for practical purposes such as most cases of GEO hazard identification. As a warning against unwarranted claims of familiarity, Regal (1993) has referred to previous bad experiences: "... the accumulated history of deliberate introductions of exotic species is a model of institutional dynamics that illustrates how dangerous it can be to assume that one is sufficiently familiar with an organism to make predictions when the familiarity is not based on a detailed understanding of the mechanisms of adaptation and range of latent adaptive potentials of the organism" (Regal 1993: 233). The general lesson to be learned here is that uses of the familiarity criterion must be based on scrupulous scientific methodology in the form of specified SRQs.

A Dutch NGO, the *Netherlands Society for Nature and Environment* (NSNE), for example, "perceives an inadequate input of up-to-date ecological insights into the decision-making on the use of GMOs [=GEOs]" (NSNE 1997: 4). A practical way of incorporating this and other criticisms in the larger procedure is by reconstructing the critique in terms of calling attention to possibly missing relevant questions. If criticism does not imply a different SRQ for the purpose of GEO biosafety assessment, then it is difficult for administrators and policymakers to appreciate its urgency.

This implies that the situation in biosafety regulation can be improved upon by accumulating experimental evidence and by careful methodological interpretation. As Ingham *et al.* (1995) have expressed it: "With so little background information available, the principle of familiarity cannot be evoked, because the background database on effects does not yet exist. But, once a number of organisms have been tested for effects, and the database developed, the principle of familiarity could be invoked as long as the temptation to extrapolate information across dissimilar ecosystems is resisted" (Ingham *et al.* 1995: 25).

To resist this "temptation to extrapolate" we must develop a scientific basis for an *adaptive* familiarity criterion in the form of a carefully designed list of relevant empirical questions to address for the purpose of GEO hazard identification. This list will never be 'complete' or 'finished', since our knowledge of and insights in biological complexity will keep developing. This implies that the level of 'familiarity' will also develop and must be kept under permanent scientific scrutiny. In this sense, the larger process of GEO hazard identification is never finished.

It will probably always be possible to think up more questions then can be answered. The importance of articulating 'familiarity' in terms of relevant research questions is that *politicians can make choices which are genuinely scientifically informed.* By making the methodological basis for claiming familiarity as explicit as possible in the form of relevant research questions, we are in the best position to distinguish the science from the politics of familiarity: "The familiarity concept also relates to risk normalization, *i.e.* comparison and categorization of risks posed by released GEMs to those posed by current practice" (van Elsas 1995: 125).

An 'Independent group of scientific and legal experts on biosafety' (1996) has recently made the following assessment: "Given the present knowledge base that is too limited, the use of the principle of familiarity in environmental situations is dangerous" (Independent Group 1996: 29). As yet, we do not have enough understanding of the biological mechanisms involved to specify the relevant research questions for GEO hazard identification. Not making the notion of familiarity *specific* enhances the chance of erring on the unsafe side in the future. The vagueness of unspecified familiarity may tempt us to decide on an *intuitive* basis that we have "enough information" to argue that an intended GEO application is "essentially similar" to previous safe experiences.

There is no way to give 'familiarity' a scientific interpretation, without doing justice to the *dynamic role of controversy*. The SRQ approach can accommodate the role of controversies in policy and administration by giving a procedural and modular interpretation of 'familiarity'. It is a method to take controversies seriously (cf. Moser 1995). Controversy is an inherent part of science. If it does not get a place in scientific expertise, the expertise will loose its scientific legitimation. The danger of being restricted to "armchair hazard identification" (NSNE 1997: 4, cf. Crawley 1994) has been warned for. One safety net for this danger is in appreciating and incorporating the persistent concern of non-governamental organizations (NGOs) and other critical parties.

— 5.5 A *dynamic* and *modular* catalog of familiarity —

The *World Bank* (Doyle and Persley 1996) has also recognized a role for the "central concept of familiarity" in biosafety policy and has emphasized, "the need for a process to be put in place by which national biosafety guidelines are continually updated to take into account learned experience and to identify new potentially hazardous processes that may result from the research process" (Doyle and Persley 1996: 21).

Since controversy and discussion is so important in this context, a practical approach should enable us to accommodate this. Therefore I propose to adopt a *dynamic* and *modular* approach to the interpretation of the familiarity criterion, that can be made responsive and adaptive to changing insights (cf. Regal 1995b). The suggested approach is *dynamic* in the sense that it is designed to develop with new insights and it is *modular* in the sense that it can be adapted and changed in small portions of individual relevant questions. One advantage of a modular approach is that it provides room for criticisms which are relatively restricted in their specific concerns. This is an important advantage because NGOs sometimes feel less prepared to offer their views because of the "difficulty of technical details" (cf. NSNE 1997).

The national *Competent Authorities* which have the task to oversee the regulation of GEO applications, can find a methodological basis for their difficult task in the development of sufficient SRQs for the purpose of GEO hazard identification. To rationalize notification procedures these SRQs should be made as explicit as possible. The questions to be included in a dynamic and modular catalog of familiarity come with a burden of proof for justifying their relevance. It would be impractical and unrealistic to include any hint or indication of a possible problem as a matter for regulatory concern. *Given* this developing catalog of relevant questions to be addressed in GEO hazard identification, the burden of proof for *excluding* specific questions listed in this catalog as not relevant from the notification procedure of a specific intended GEO use lies with the *Competent Authority*.

Before drawing the contours of such a catalog of possibly relevant research questions to be included in SRQs for specific research purposes (Section 5.6), I will first demonstrate why most of the present approaches to the use of the familiarity criterion in notification procedures do not satisfy the purposes of biosafety assessment. I will do so by discussing the *European guidelines* (Section 5.5.1) for notification and the *Dutch guidelines* for notification of intended GEO use (Section 5.5.2).

5.5.1 Annex II of EU council directive 90/220

In "An appraisal of the working in practice of directive 90/220/EEC on the deliberate release of genetically modified organisms", Von Schomberg (1998a) has criticized the role of scientific expertise and the appeal to science in the current practice: "The appeal to science for policy has been made without reflecting sufficiently on the transscientific issue underlying a scientific controversy. Science *reduced* this issue by translating it into a question of *relevancy* to which both molecular biologists and ecologists came up with unsatisfactory answers. As a consequence, the contradiction that arises between policy and science has not been reflected either. Policy has to be engaged in science to look for answers concerning perceived risks but cannot make a legitimate appeal to a science which does not resolve the 'relevancy' question" (von Schomberg 1998a: 6-7 – no italics added; cf. von Schomberg 1998b).

Whereas I agree with the general diagnosis of this appraisal that the "relevancy question" has not been resolved, I disagree with the suggestion that science *could* not resolve it. A methodological analysis in terms of SRQs can provide us with the methodological boundaries of scientific legimation for biosafety assessment. In my view, this cannot be replaced by any form of political legitimation. On the other hand, the *acceptability* of uncertainties involved with scientific hazard identification should indeed be subjected to political legitimation and cannot be resolved in a scientific arena. Therefore, I conclude that a scientific interpretation and definition of the familiarity criterion remains a *conditio sine qua non* for a democratically based discursive procedure on the acceptability of uncertainties in the scientific identification of possible GEO hazards.

My criticism is in agreement with the concern expressed by Von Schomberg about the, "absence of a definition of environmental harm" (von Schomberg 1998a: 5; cf. von Schomberg 1996, 1997). The only practical solution that I see to address this concern is by explicitly specifying the relevant research questions that need to be asked in order to recognize 'harm' or 'hazard' (for more detail, see Section 5.6). There is no more direct way to discuss 'hazardous' effects than by asking how they can be detected or recognized. As it is, the information requirements of EU *Annex II* are only seemingly 'empirical'. *Annex II* rightly states, for example, that, "[t]he level of detail required in response to each subset of considerations is (...) likely to vary according to the nature and the scale of the proposed release" (CEC/*Annex II* 1990: 23), but it does not specify the proper scientific way to "vary" this "level of detail required". A practical approach in terms of relevant research questions can help us accommodate this requirement.

In their present form the "information requirements" as specified by *Annex II* of *Council Directive* 90/220 leave open too much freedom to the applicant for interpretation of the notification requirements, which is not in the interest of a scientific biosafety assessment. In other words, the "information required in the notification" does not sufficiently specify the relevant research questions. It presumes a superficial interpretation of the notion of 'familiarity'. The present level of specification in *Annex II* takes the form of vague and informal phrases such as "information on survival ...", "nature and source of the vector ...", "predicted habitat of the GMOs ...", "likelihood of post-release selection ..." (CEC/*Annex II* 1990: 23-27). All of these "information requirements" address complex biological processes which can only be studied by using sufficient detail.

Such vague terms as listed in *Annex II* of EU *Directive* 90/220 enhance the methodological risk that the notion of 'familiarity' will be applied in its everyday interpretation instead of in a rigid scientific definition. It also makes it virtually impossible to call a specific notification 'insufficient' on the basis of these requirements. This is not just a bureaucratic concern, it touches upon the very kernel of an adequate hazard identification for the purpose of GEO biosafety assessment. The approach taken to 'familiarity' in *Annex II* is therefore unsatisfactory for its purpose.

5.5.2 The Dutch advisory board COGEM

To discuss a more specified scientific approach to the information requirement for the notification of intended use of a GEO, the working practice of the Dutch advisory board COGEM (*Commissie voor Genetische Modificatie*) will now be analysed as an example. How does the COGEM deal with the "relevancy question" of scientific information? In my view, the COGEM lacks a procedural methodological approach for dealing with possibly relevant research questions and therefore cannot accommodate a scientific interpretation of the familiarity criterion.

In a Dutch newsletter of professional biologists, a *safety officer* of the *Agricultural University of Wageningen* appealed to the readers to discuss the scientific basis of guidelines for the use of genetically engineered microorganisms in laboratories (Middelhoven 1997). Middelhoven specifically raises the issue of imperfect physical containment of uses of genetically engineered *Agrobacterium tumefaciens* and invites the Dutch community of biologists to discuss the question whether we have sufficient knowledge of the *microbial ecology* of genetically engineered microorganisms such as *A. tumefaciens* to legitimate creating the possibility of its unintended release into the environment.

Middelhoven (1997) argues that, in his view, specific research questions are relevant for the purpose of GEO hazard identification. The response of the COGEM to this appeal by voice of its Chairman and its Secretary is less than satisfactory from a scientific point of view (Schellekens and Bergmans 1997). Where the COGEM representatives decide not to take on the *scientific* burden of proof for excluding the possibly relevant research questions as suggested by Middelhoven (1997) from a sufficient SRQ for the specific research purpose, they thereby take on the *political* burden of proof for arguing the acceptability of using a less than scientific basis of legitimation for their review of notifications.

With a more dynamic and adaptive approach to the familiarity criterion, the COGEM could perhaps have responded in a more receptive and scientific vein. As it is, in their initial response the Chairman and the Secretary responded only to a minor part of the concern expressed by Middelhoven. They countered the criticism by saying that the *physical containment* of a laboratory or test site cannot be perfect and that we should accept this limitation. However, the challenge for debate that was posed by Middelhoven (1997) referred to our familiarity with the *biological containment* of genetically engineered microorganisms such as *Agrobacterium tumefaciens*. In my view, the scientific legitimation of an advisory board such as the COGEM should imply that they let their advice benefit from the input of concerned professionals such as Middelhoven and the community of Dutch biologists.

On second thoughts the COGEM did take up the concern expressed by Middelhoven as food for consideration. Deliberation about the issue among the Dutch *Ministry of the Environment*, the COGEM, and the *Agricultural University of Wageningen*, led to a joint publication in *BioNieuws* in which the Secretary of the COGEM, Bergmans, and Middelhoven expressed a "common point of view". They conclude with the statement that, "[t]he accurate compliance with the advised code of conduct will sufficiently prevent the possibility of introduction [of genetically engineered *Agrobacterium tumefaciens*] into the environment" (Bergmans and Middelhoven 1998: 2 – transl. AVD). Besides, the joint statement specifies that the choice of, "method or combination of methods applied is left to the circumstances and the judgement of the researcher and supervising biological safety officer" (Bergmans and Middelhoven 1998: 2 – transl. AVD).

The discussion has shifted its centre of focus along the road, which is regrettable from a scientific point of view. Where Middelhoven initially raised concern about our familiarity with the *biological containment* of *A. tumefaciens* and the COGEM responded unsatisfactorily by only addressing questions of *physical containment*, now, in the joint viewpoint, the concern about sufficient knowledge of biological containment has completely disappeared from the argument. The *answer* to the initial question is restricted to a specification of measures to take for sufficient physical containment. However, the question about the biological containment of transgenic *A. tumefaciens* has not been (and cannot be) addressed along these lines. Besides, in other discussions the concern has been raised that 'physical containment' and 'biological containment' should not be treated as 'absolutes' but rather as on a continuum of containment (cf. Wheale and McNally 1996: 188; see also Section 4.5). Thus, the conclusion must be that the *relevance* of the question on biological containment has not been denied *and* the question has not been sufficiently addressed either.

There is a lot at stake in terms of *trustworthiness* for an advisory board also. It would be highly regrettable and unproductive if an image would develop of advisory boards at the assistance of *Competent Authorities* such as expressed by one representative of *Friends of the Earth* in relation to the UK *Advisory Committee on Releases to the Environment* (ACRE): "... [ACRE] has thirteen members. Eight of them have direct ties with biotechnology companies. One member could be characterized as 'green'. A permit application has never been denied. It is as if wolves are herding the sheep" (cit. in Wagendorp 1998 – transl. AVD). In my view, the only remedy against this kind of negative image-building in relation to genetic engineering advisory boards, is to develop and apply a scientific approach to discussions about familiarity. This is in the best interest of biotechnology companies, as well as of *Competent Authorities* and of societies in general.

The Secretary of the Dutch advisory board COGEM does acknowledge the necessity of giving a scientific and adaptive basis to the use of the familiarity criterion: "This requires that the hazardous situation is compared to the actual situation in the actual environment; knowledge about these 'base lines' is not readily available..." (Bergmans 1995: 23). To demonstrate how fruitful a procedural interpretation of familiarity can be in practice, I give some more examples of how the COGEM has been offered assistance from critical commentators such as concerned professionals and NGOs.

One Dutch NGO, the *Netherlands Society for Nature and Environment* (NSNE), took the initiative to assemble an "Inventory of reviews of non-governmental organizations on the risk evaluation of genetically modified organisms in six cases" (NSNE 1997). Several NGOs were invited to give their view on whether the level of familiarity at the basis of the notifications and the permits was sufficient for the purpose. They were asked to respond to such questions as: "Do you consider the data submitted to characterize the host organism as adequate? If not, why not? Which additional data should be provided in your view, and why these data?" (NSNE 1997: 67). The responses can be seen as potentially valuable input in a learning process towards greater familiarity with GEO hazard identification. In my view, it is up to the COGEM to be the coordinator of this learning process.

NGOs which raise a research question as relevant for a specific notification thereby take on the burden of proof to argue for this relevance. Controversies about different cases can be expressed in a scheme such as given in Figure 5.2, in which the antagonists have to provide arguments for claiming the relevance of a particular research question. If the COGEM (or the *Competent Authority*) does not want to consider the questions cited as relevant, then the scientific burden of proof for their *ir*relevance in this context comes to lie with the COGEM.

Many more questions have been and are being raised by, in many cases highly scientifically qualified, representatives of NGOs (cf. Rissler and Mellon 1996; Independent Group 1996). It seems to me that a *Competent Authority* should either address these questions or argue why they can be reasonably excluded from a sufficient SRQ for a specific GEO notification.

What the COGEM needs, as do other advisory boards and the *Competent Authorities* they serve, is an *evaluative framework* in which these and other research questions raised by the efforts of NGOs and others can be considered for their *relevance* in relation to the specified research purpose of GEO hazard identification. In so far as such critical assistance from NGOs and others is not made productive for the general purpose of biosafety assessment, it is a waste of expertise both from the point of view of science and from the point of view of society.

Question raised by *safety officer* "Can we rely on the *biological containment* of transgenic *Agrobacterium tumefaciens*?" (Middelhoven 1997)	Argument for *relevance* "Available knowledge of *microbial ecology* is insufficient"
Question raised by NGO "Which criteria have been used in this case to establish whether the host organism is non-pathogenic?" (NSNE 1997: 67)	Argument for *relevance* "Pathogenicity of (transgenic) organisms may vary with varying biological context"
Question raised by NGO "Are there data on the possible exchange of genetic material between the strains and autochtonous *Pseudomonaceae* available?" (NSNE 1997: 68)	Argument for *relevance* "Transfer of transgenes between GEOs and wild-type strains may have inadvertent consequences"
Question raised by NGO "Do the strains under notification indeed have a lower survivability than the natural isolate Ps130?" (NSNE 1997: 72)	Argument for *relevance* "Competitive ability of GEOs is relative to other organisms"
Question raised by NGO "Why are no data included on the number of integration sites and the number of copies per integration site" (NSNE 1997: 88)	Argument for *relevance* "Number of transgene copies and sites may affect expression of transgenic trait and other phenotypic characteristics of GEO"

FIGURE 5.2 Schematized representation of possibly relevant research questions raised by NGOs (NSNE 1997) and other critical parties interpreted as an attempt towards a sufficient set of relevant questions for the hazard identification of GEO release in the format of a procedural and modular approach. An assumed SRQ can be adapted and changed with the precision of individual questions to be included in or excluded from the 'window of concern'. The questions in this example represent some of the controversies analysed in this study. For the present purpose, the *procedure* of choosing relevant research questions is more important than the particular research questions discussed. The right-hand column represents the mechanism of selecting the questions in the left-hand column and arguments for their (ir)relevance.

This approach to the familiarity criterion would help us give a scientific basis to the information requirements of notifications for intended GEO use. A dynamic and modular approach to the science of hazard identification shows how the screening efforts of NGOs can provide valuable assistance as helping hands or 'hazard detectives' to the work of the *Competent Authorities.* A question raised (and argued) as possibly relevant for GEO hazard identification should either be addressed or be excluded from concern on the basis of scientific arguments for it's not being relevant; abstaining from either does not seem to be a methodologically justifiable option.

— 5.6 Towards a sufficient set of relevant questions —

In a report by the *World Bank* (Doyle and Persley 1996) it is argued that, "[t]he concept of *familiarity* can assist decisionmakers in evaluations by providing a context in which to apply accumulated experience..." (Doyle and Persley 1996: 12). To live up to these expectations we need a practical procedure to make this "accumulated experience" accessible for its users.

An extensive reservoir of highly qualified scientific literature is available on GEO hazard identification, but does not have its potential impact for lack of an integrating framework. In many theoretical and empirical studies, possibly relevant research questions are raised and addressed. In the present situation it remains too easy for antagonists to dismiss contested biosafety claims and put forward more convenient alternative answers. The best chance for arriving at a robust and developing catalog for the familiarity criterion is to bring together those possibly relevant questions and make them the centre of world-wide scientific discussion and research.

Many participants in the biosafety debate have produced candidate "checklists" of safety concerns for GEO hazard identification (cf. Barnes and Hulsman 1995; Charudattan 1990; CEC 1990; Doyle and Persley 1996; van Elsas 1995; Käppeli and Auberson 1997; McCammon and Medley 1990; Miller 1988; NAS 1989; Norton *et al.* 1992; OECD 1986; Plant 1990; Powell 1995; Rissler and Mellon 1993; Seidler 1994; Tiedje 1990; Tiedje *et al.* 1989; UNEP 1996; Zilinskas 1995). These and other studies can be used as *resources* for the development of a sufficient SRQ for a specific GEO hazard identification; they can be seen as *supplementing* and *complementing* 'windows of concern' from which a common perspective can be developed for particular purposes. The SRQ approach allows us to bring together the already available insights and information in a constructive way, without causing paradigmatic controversy.

An international research project for developing a scientific definition of 'familiarity' for the purpose of GEO hazard identification could become a valuable tool for making biosafety controversies productive. Up-to-date versions of a developing *catalog of familiarity* and world-wide research results on GEO hazard identification could be made globally available to *Competent Authorities* and other concerned parties via the *Internet.* Such an integrative *web site* for biosafety assessment could become an international platform for scientific discussion on the relevance of research questions for GEO hazard identification (see also Section 2.6). Part of a global effort to define 'familiarity' for GEO hazard identification, could be to promote the accessibility of data banks via *Internet* (cf. Schmitt 1995: 220).

An important concern for biosafety *management* is to find ways to incorporate critical expertise. Most existing "checklists" do not do justice to the role of controversies in scientific expertise. In a SRQ approach, controversies can be incorporated as a potential source of progressive fine-tuning of our understanding and familiarity. By stressing the importance of the burden of proof in the methodological justification of scientific claims, we have opened up an avenue for appreciating the critical input of NGOs and other "orthodox or unorthodox experts" (cf. Lake 1991). Along these lines, controversies can become an *asset* rather than an *obstacle* towards more and more adequate "flowcharts" or SRQs for biosafety testing. The *development* of science is not sufficiently accommodated in most checklists. To remedy this omission, I plead for a *procedural* and *modular* approach to 'familiarity' as the basis of a world-wide project towards a sufficient *set of relevant questions* for the purpose of GEO hazard identification.

As a representative attempt towards "Enabling the safe use of biotechnology" consider the following approach to GEO hazard identification: "Important factors in risk assessment are the degrees of hazard posed by: a. the parent (wild type) organism; b. genetic constituents – donor DNA; c. phenotype of organisms with novel traits; d. attributes of the environment" (Doyle and Persley 1996: 22-23). This general format for familiarity does not adequately address the relevant *categories of questions.* The general approach resembles the one that was discussed in Chapter 3 in relation to the controversy between the possibility of finding the possible hazards of a GEO by "adding up" the hazards of the individual elements from which the GEO was 'assembled'. Critics have observed that this approach rules out the possibility of considering 'synergistic' effects.

By considering the *donor*, the *recipient*, the *vector*, the *transgenic insert*, and the *environment* separately, there is not a sufficient scientific basis for the claim that 'familiarity' with the *parts* will provide sufficient familiarity with the new *ensemble.* Besides, such an "adding up" approach to GEO hazard identification does not give us a practical basis for the *monitoring* of biosafety experiments or for doing *field tests*. Since the constituting 'parts' have 'disappeared' into the whole of the new GEO and its release, the cited approach to "degrees of hazard" will not be functional in a monitoring process. This approach lacks a methodologically adequate operationalization of the concept of 'GEO hazard' (cf. von Schomberg 1998a: 5).

The following operationalization of the concept of 'GEO hazard' may provide a practical methodological basis for a scientific interpretation of 'familiarity'. Its adaptive structure allows for improvement in the course of its use:

> A 'GEO poses a *hazard*', if 'it carries an *Agent P*, which can impose an *Effect Q* which is considered undesirable in *Context R*, to an *Affected S*, via a *Mechanism T*, in an *Environment X*, as a consequence of *Application Z*'.

The separate ingredients of this definition of 'GEO hazard' can be seen as *categories of possibly relevant questions* which together constitute a *catalog of familiarity* for GEO hazard identification from which a sufficient SRQ for a specific GEO notification can be selected (see Sections 5.6.1-5.6.7 and Figures 5.3-5.9 – not making any claim to completeness).

To make biosafety claims without specifying the 'agent(s)', 'effect(s)', 'context(s)', 'affected', 'mechanism(s)', 'environment(s)', and 'application(s)' that could be involved with a possible GEO hazard is a violation of the scientific burden of proof for justifying empirical claims in the context of GEO hazard identification. Notifications in which these constituting elements of possible hazards are not addressed as relevant concerns therefore *lack a sufficient scientific basis of legitimation*. This does not take away the possibility that the applicant or the *Competent Authority* may produce scientific arguments for *not* including specific research questions in the SRQ for a specific GEO hazard identification.

This approach to GEO hazards necessarily includes a *redundancy* in hazard considerations, since the categories of relevant questions are not *independent*. For example, considerations about a 'mechanism' of GEO hazard may very well immediately imply considerations about 'effect(s)' of GEO hazard. However, redundancy does not imply superfluousness here. A concern that may be missed in one category of relevant questions, may turn up as relevant in another category of concern. Thus, this redundancy of the categories of concern may be seen as an advantage rather than as a burden for the process of GEO hazard identification.

The presented approach also does justice to the fact that, for practical purposes, it is very difficult to separate different concerns such as "ecological and genetic considerations" (OTA 1988). The two can be *distinghuished* in research, but are virtually *inseparable* in biological systems. An important additional advantage of this operationalization of GEO hazardousness is that it offers specific clues for the design of GEO *monitoring* programmes.

5.6.1 Agent(s) of GEO hazard?

To recognize or detect a hazard, insight is required in the relevant questions to ask in relation to the *agent(s)* of a GEO hazard. This may not always be a straigth-forward challenge. One angle from which it can be approached is to question the characteristics of the constituting elements of the GEO under notification. Are the *vector*, the *donor*, the *recipient*, or the *genetic insert* known to be a cause for concern? These are important questions to address, but they do not constitute a *sufficient* set of relevant questions.

One additional concern is that the GEO itself may have characteristics that exceed the collection of assembled characteristics. Therefore we must also consider the question about possibly relevant characteristics of a GEO. For example, what are the *gene products* of the transgene in the GEO? This is not necessarily limited to the already known products of the constituting parts of the GEO. Also, can an affected actor on the biological scene become a hazardous agent itself, such as in the case of the rhizosphere that was affected by transgenic *Klebsiella planticola* and thereby 'indirectly' became a hazardous agent for the wheat plants (see Section 2.2.1)?

Other relevant questions may pertain to the *fate of the transgene* or of the *naked* DNA, which may turn out to be a hazardous agent by itself. Can the transgene survive outside the GEO? If so, under what circumstances? How stable will the transgene be incorporated in the genome of the GEO? How many copies of the transgene will be in the GEO? Will the number of integrated copies affect the GEO characteristics? Can transposons become active hazardous agents? Maybe these questions need not all be considered as relevant in the review of all GEOs, but then reasons must be given in those cases why these questions can be excluded from consideration as irrelevant.

It is important to note that such a list of possibly relevant questions about agent(s) of GEO hazard does not imply any claim to *completeness*. The applied notion of 'familiarity' will (have to) change with developing insights (see Figure 5.3). This means that questions considered relevant now may be taken off the list later and it may also mean that new questions will be added to the list in the future as possibly relevant for consideration. The first type of change requires argument for the *irrelevance* of the excluded question(s), the latter type of change to the catalog of familiarity requires argument(s) for the *relevance* of the included question(s).

Possibly relevant questions about *agent(s)* of GEO hazard	
Agent of GEO hazard? Are the *vector*, the *donor*, the *recipient*, or the *genetic insert* known to be a cause for concern?	Argument for *relevance* Previous experiences may be indications of future hazards
Agent of GEO hazard? What are the *gene products* of the transgene in the GEO?	Argument for *relevance* New gene products may constitute new hazardous agents
Agent of GEO hazard? Can the *transgene* survive outside the GEO?	Argument for *relevance* The transgene may become a hazardous agent by itself
Agent of GEO hazard? Will the *copy number* of an integrated transgene affect the GEO characteristics?	Argument for *relevance* The *number* and *place of integration* of transgenes in the genome may affect the traits of the GEO
Agent of GEO hazard? ... other relevant questions...?	Argument for *relevance* ... include in SRQ...?

FIGURE 5.3: Schematized representation of possibly relevant questions on the *agent(s)* of GEO hazard in a procedural and modular format for a catalog of familiarity from which a selection can be made to arrive at a sufficient *set of relevant questions* for the hazard identification of a particular GEO release. An assumed SRQ can be adapted and changed with the precision of individual questions to be included in or excluded from the 'window of concern' for notification. For the present purpose, the *procedure* of choosing relevant research questions is more important than the particular research questions discussed. The right-hand column represents arguments for their relevance (no claim for *completeness* implied).

5.6.2 Effect(s) of GEO hazard?

The notion of 'effect' is obviously central to GEO hazard identification. An important concern is that effects may take many guises. What is a *desirable* effect in one place and time, may be or become *undesirable* in other places and times. Also, as was shown for the case of transgenic *Klebsiella planticola*, effects may be hidden from immediate sight and may challenge the 'sensory organs' of science to make them 'visible'. For example, what will be the effect of a transgenic *gene deletion* on a GEO? Will it always result in the loss of (a) trait(s)? And if so, could that loss of trait(s) pose a hazardous effect itself?

A general concern is that impacts which may seem to be 'side' effects, may turn out to be important upon closer scrutiny. For example, if transgenic herbicide-resistant crops allow the use of herbicide in more sensitive growth stages (cf. Krimsky and Wrubel 1993), could this have the *side* effect that a higher *residue* of the herbicide will be consumed with the GEO? And if so, could this affect the food safety for its consumers? What may seem to be 'side' effects from one perspective may turn out to be central concerns from another perspective (cf. Bell 1999). An example of this general concern can be found in the relevance of more specific research questions, such as: "... cumulative and non-linear effects are strangely enough completely neglected in risk evaluations of genetic modifications" (Reijnders 1992: 101).

Van Kuijen (1992) rightly notes that, in the context of GEO hazard identification, "[t]he word effect is used in many different ways". In his interpretation, possible effects can be, "the very first effect of a new gene in an organism, the production of a gene product", "the effect of the gene product on an organism", "the effect of the organism with the new trait on its direct neighbours", "the long term effect of an organism which influences a certain ecosystem because of its excessive growth", "the effect of the new gene of transferred DNA to other organisms" (van Kuijen 1992: 25). Making these distinctions implies that the relevance of the related research questions must be addressed and be made part of a *catalog of familiarity* as possible reason for concern (see Figure 5.4).

Possibly relevant questions about *effect(s)* of GEO hazard

Effect of GEO hazard? What may be effects of *gene deletions*?	Argument for *relevance* *Loss of trait(s)* may also change the impact of a GEO on other organisms or ecosystems
Effect of GEO hazard? Can new regimes of pesticide use lead to higher *pesticide residues* in GEOs?	Argument for *relevance* Residues of pesticides in transgenic crop plants may affect food safety
Effect of GEO hazard? What consequences for biosafety may result from *loss of genetic diversity*?	Argument for *relevance* Transgenic crops may have less genetic variation than traditional crops
Effect of GEO hazard? What consequences may result from *hybridization* of a GEO with wild-type species?	Argument for *relevance* Outcrossing of transgenic trait(s) may create new pests or disturb natural ecosystems
Effect of GEO hazard? ... other relevant questions...?	Argument for *relevance* ... include in SRQ...?

FIGURE 5.4: Schematized representation of possibly relevant questions on the *effect(s)* of GEO hazard in a procedural and modular format for a catalog of familiarity from which a selection can be made to arrive at a sufficient *set of relevant questions* for the hazard identification of a particular GEO release. An assumed SRQ can be adapted and changed with the precision of individual questions to be included in or excluded from the 'window of concern' for notification. For the present purpose, the *procedure* of choosing relevant research questions is more important than the particular research questions discussed. The right-hand column represents arguments for their relevance (no claim for *completeness* implied).

5.6.3 Context(s) of GEO hazard?

In some contexts, a specific effect such as *pathogenicity* or *toxicity* will be a wanted effect, *e.g.*, for 'fighting' insects. In the language and the practice of methods in agriculture its daily business is depicted as being about 'warfare' in part. A *war* against disease, a *war* against insects, against drought, etc. The kernel of hazard identification in genetic engineering for agriculture is that we do not want this 'war' to backfire on ourselves.

For example, even *absence* of an expected effect such as herbicide-resistance can lead to severely detrimental effects (cf. Broer 1995). It has been shown that herbicide-resistance may not be expressed by a transgenic plant when the temperature is too high. This could mean that when herbicides were used on a hot day, the whole crop could suffer despite its transgenic herbicide-resistance. Apart from economic cost, in some areas of the world this could lead to famine also.

In many cases, (un)desirability of an effect will depend upon the specific *context*. For example, *tolerance* for frost or drought or flooding may be desirable in one context and undesirable in another context of use (see Figure 5.5). Also, concerns have been raised about the use of inserted *markers for antibiotics resistance* in GEOs for research purposes. In a medical context these 'markers' may become a problem by obstructing the effective use of those antibiotics (cf. Davies 1994). Thus, the demands that seem to come naturally with the context of research and of agriculture, may turn out to be conflictuous with the demands that are primary in the context of medicine and of food safety (cf. Teuber 1996; Nordlee *et al.* 1996). Depending on the context of use, new or other research questions may become relevant for the purpose of GEO hazard identification.

Possibly relevant questions about *context(s)* of GEO hazard

Context of GEO hazard?	Argument for *relevance*
Is a specific GEO effect also desirable in other contexts of use?	Intended or unintended use in other contexts may lead to new and possibly undesirable effects
Can plants with *herbicide resistance* become a weedy nuisance in another context?	Plants which are a *crop* in one context may be considered a *weed* in another context
Will *tolerance* of a GEO for *drought*, *flooding*, or *frost* be desirable in other contexts of use?	Specific GEO *tolerances* may lead to a competitive advantage and to inadvertent colonization
Can *genetic markers* have an undesirable biological function in another context of use?	Loss of usefulness of specific antibiotics by the application of *antibiotics markers* in GEOS
... other relevant questions...?	... include in SRQ...?

FIGURE 5.5: Schematized representation of possibly relevant questions on the *context(s)* of GEO hazard in a procedural and modular format for a catalog of familiarity from which a selection can be made to arrive at a sufficient *set of relevant questions* for the hazard identification of a particular GEO release. An assumed SRQ can be adapted and changed with the precision of individual questions to be included in or excluded from the 'window of concern' for notification. For the present purpose, the *procedure* of choosing relevant research questions is more important than the particular research questions discussed. The right-hand column represents arguments for their relevance (no claim for *completeness* implied).

5.6.4 Affected of GEO hazard?

Biological complexity implies that it is difficult to predict what will be affected. As Sterling (1990) notes: "There are few technologies that impact only the target component" (Sterling 1990: 154). For a general impression of the complexity at hand, consider Sterlings perspective on "non-target" biological components: "The term 'non-target population' evokes different conceptual images depending on one's experience. The environmentalist may see endangered species, the entomologist natural enemies, and the molecular biologist cellular, hormonal, or immunological systems" (Sterling 1990: 153). Also, the possibly affected may become weakened and thereby more susceptible due to less direct impacts such as loss of biodiversity and/or environmental pollution.

As an example of how a concern for "non-target" impacts may be relevant for GEO hazard identification, consider the following example: "Although *Bacillus thuringiesis* has been used safely to combat leaf-eating *lepidoptera* for many years, the new product is designed to affect soil organisms. The shift of a gene from a leaf-dwelling to a soil-dwelling bacterium will expose many other soil organisms to a toxin that they have quite likely not encountered before in significant quantities in their own microenvironment. Both new target organisms and beneficial nontarget species could be affected" (Sharples 1991: 21). This illustrates how concerns about the possibly affected must be raised as a standard procedure in GEO hazard identification (see Figure 5.6).

Possibly relevant questions about *affected* of GEO hazard	
Affected of GEO hazard? How is the affected *target* distinguished from possible *non-target* affected?	Argument for *relevance* Non-target affected may share susceptibilities with target affected
Affected of GEO hazard? Will an impact have consequences for a *limited number* of affected?	Argument for *relevance* A hazard posed to *individuals* rather than *populations* may be less problematic
Affected of GEO hazard? Can the transgene(s) change the affected *host range* of a GEO?	Argument for *relevance* Change in the *host range* may affect non-target species
Affected of GEO hazard? What is the *secondary* effect of change to the target on *non-target* affected?	Argument for *relevance* New pests may get a chance to develop in the absence of a target pest
Affected of GEO hazard? ... other relevant questions...?	Argument for *relevance* ... include in SRQ...?

FIGURE 5.6: Schematized representation of possibly relevant questions on the *affected* of GEO hazard in a procedural and modular format for a catalog of familiarity from which a selection can be made to arrive at a sufficient *set of relevant questions* for the hazard identification of a particular GEO release. An assumed SRQ can be adapted and changed with the precision of individual questions to be included in or excluded from the 'window of concern' for notification. For the present purpose, the *procedure* of choosing relevant research questions is more important than the particular research questions discussed. The right-hand column represents arguments for their relevance (no claim for *completeness* implied).

5.6.5 Mechanism(s) of GEO hazard?

Understanding and detection of possible hazardous effects may be improved by sufficient awareness of relevant biological *mechanisms*. Many natural mechanisms have been cited as possible cause for concern: *epistasis*, *pleiotropy*, *polygeny*, *transformation*, *transduction*, *conjugation*, etc. More ecological concerns are *adaptation*, *spread*, *invasiveness*, *lateral gene transfer*, etc. The *catalog of familiarity* could serve as a heuristic checklist of possible concerns in relation to specific biological mechanisms that could be particularly relevant for GEO hazard identification (see Figure 5.7).

Study of the biological mechanisms that could be involved with GEO hazards is obviously important, and at the same time very difficult. Many variables may be involved, and even for seemingly straightforward impacts on an ecosystem it may be quite a challenge to recognize the relevant system objects, the relevant system relations, and the relevant system dynamics. To gain insight in the possibility for a GEO to expand its ecological range and to evade a new community, several 'separate' factors can be involved: "the availability of resources, pathogens, herbivores, seed predators, competition with other plant species, the availability of symbionts such as mycorrhizal fungi, or the availability of open sites for germination and seedling establishment" (Schmitt and Linder 1994: 71).

To give a sufficient methodological basis to biosafety claims on the impact of GEO releases, more 'familiarity' is required with the dynamics and stability of an ecosystem: "To assess the ecological risk of a particular transgenic manipulation, it is therefore critical to know what currently limits populations of the target plant species and its wild relatives, as well as how the genetic modification will affect those limits" (Schmitt and Linder 1994: 72). For lack of readily available ecological theory, the practical way to proceed is to 'sum up' the possibly relevant research questions and thereby get a better grasp on the problem of investigation. This effort could in itself contribute to scientific progress with respect to insight in relevant biological mechanisms such as *invasiveness*, because so far: "Unfortunately, surprisingly little is known about ecological limits on the distribution and abundance of plant species" (Schmitt and Linder 1994: 72).

One 'factor' in plant invasiveness that is relatively well-known but still of unclear relevance to GEO hazard identification has been pointed out by Reijnders (1992). In his view, the spread around the globe of agricultural activity itself, "invasions by agricultural protection" (Reijnders 1992: 101), could be a cause for concern.

Possibly relevant questions about *mechanism(s)* of GEO hazard

Mechanism of GEO hazard?	Argument for *relevance*
Will there be adaptative ecological effects over longer periods of time?	Ecosystem change may only take effect over a longer period of time
Can *transposon activity* pose a hazard for GEO release?	Transposons may affect the *genomic stability* of a GEO
Can *lateral transfer* of transgenes pose a hazard?	'Escape' of transgenes could create new pests
Weedy invasiveness through available space in recipient ecosystem?	Different ecosystems may have different susceptibility to invasiveness of GEOs
... other relevant questions...?	... include in SRQ...?

FIGURE 5.7: Schematized representation of possibly relevant questions on the *mechanism(s)* of GEO hazard in a procedural and modular format for a catalog of familiarity from which a selection can be made to arrive at a sufficient *set of relevant questions* for the hazard identification of a particular GEO release. An assumed SRQ can be adapted and changed with the precision of individual questions to be included in or excluded from the 'window of concern' for notification. For the present purpose, the *procedure* of choosing relevant research questions is more important than the particular research questions discussed. The right-hand column represents arguments for their relevance (no claim for *completeness* implied).

5.6.6 Environment(s) of GEO hazard?

Different traits will have different effects in different environments. *Biodiversity* may well be considered as another expression of the great variety of habitats and ecosystems around the globe. This implies that GEO hazard identification must be concerned with the effects a GEO may have in *different* environments (see Figure 5.8).

Bergelson *et al.* (1998) have recently studied the possibility of spread of transgenes to weedy species. What are the relevant questions to address about the receiving environment? Depending on the environment of release, different concerns are be likely to be more relevant: "If the potential recipient of a transgene is a highly selfing species, such as *Arabidopsis thaliana*, th[e] risk [of transgene spread to weedy species through hybridization] is often considered negligible" (Bergelson *et al.* 1998: 25).

However, under the title "Promiscuity in transgenic plants", Bergelson and colleagues present the results, "... of a field experiment in which transgenic *A. thaliana* showed a dramatically increased ability to donate pollen to nearby wild-type mothers compared with *A. thaliana* mutants expressing the same mutant allele as the transgenic plants" (Bergelson *et al.* 1998: 25). Although the underlying genetic mechanism is not (yet) understood, these 'hazard detectives' conclude that: "..., these results show that genetic engineering can substantially increase the probability of transgenic escape, even in a species considered to be almost completely selfing" (Bergelson *et al.* 1998: 25). The findings suggest that use of transgenic organisms in natural environments may pose new challenges to our understanding of biological complexity (cf. van der Weele 1999).

Even if there is no intention of using a GEO in different *geographic regions*, this possibility may also be a relevant concern in hazard identification. A practical 'problem' is that ecosystems most often do not stop for national borders. In relation to this it may even be considered a matter of *biosafety* to consider whether or not *political stability* may be expected in a region of release or of field-testing. Geographic regions with a high incidence of flooding or storms may also require special concern in relation to the expected environment of release. Even a deliberate release could *accidentally* reach a different area than was targeted (cf. Ho and Tappeser 1997; Regal 1997).

Possibly relevant questions about *environment(s)* of GEO hazard

Environment of GEO hazard?	Argument for *relevance*
What is the *relative competitive ability* of a GEO in relation to other organisms?	GEO competitiveness is not an 'absolute' measure, but relative to its competitors
How can the impact of *biological diversity* in different geographic regions be assessed?	*Microcosm* studies will always address a limited biological diversity
Can GEOs with *drought tolerance* colonize dry areas?	Sensitive ecosystems may be affected by competing colonizers
Does the *re*introduction of a GEO in the same environment imply absence of hazard?	Even small genetic changes may have new ecological consequences
... other relevant questions...?	... include in SRQ...?

FIGURE 5.8: Schematized representation of possibly relevant questions on the *environment(s)* of GEO hazard in a procedural and modular format for a catalog of familiarity from which a selection can be made to arrive at a sufficient *set of relevant questions* for the hazard identification of a particular GEO release. An assumed SRQ can be adapted and changed with the precision of individual questions to be included in or excluded from the 'window of concern' for notification. For the present purpose, the *procedure* of choosing relevant research questions is more important than the particular research questions discussed. The right-hand column represents arguments for their relevance (no claim for *completeness* implied).

5.6.7 Application(s) of GEO hazard?

One obvious way in which the category of *'application(s)'* is relevant for the identification of GEO hazards is the 'exposure' of the environment to the GEO. One may ask, for example, whether 'biological containment' need not be considered for intended applications in 'physical containment'. A general cause for concern in relation to both physical and biological containment may be found in the following observation: "The essential point is that of time scale. *When the probability is low but not zero it will happen sooner or later.* The question then is where to put the borderline in time: in our own life, that of our children or do we take the evolutionary point of view?" (van Damme 1992: 90).

Different types of applications may carry different possible hazards. It may be, for example, that a *single* environmental release of some GEO does not pose a hazard, while a *repetitive* application (as will often be the intended case in an agricultural context) does pose a hazard. Another concern of GEO application may be the local density in which a GEO will be introduced in some environment. In some cases, a *low density* may not pose a problem whereas a *high density* does (see Figure 5.9).

A general concern of application may be that the use of genetically engineered *micro*organisms will typically imply the use of large numbers, which may pose a hazard in itself: "..., because of the large number of microorganisms that are likely to be released each time a product is used and the frequency with which the product will be used, especially following commercialization, even low-probability events may occur at an observable frequency" (Strauss 1991: 298).

Concerns about specific applications will most likely develop with the development of agricultural biotechnology (cf. Doyle *et al.* 1995: 243). New questions may arise as possibly relevant and earlier concerns may perhaps be dismissed as irrelevant when experiences and research results accumulate.

Possibly relevant questions about *application(s)* of GEO hazard

Application of GEO hazard?	Argument for *relevance*
Does the *local concentration* of GEOs in agricultural applications have specific impact?	'Adaptive valleys' or 'evolutionary thresholds' may be overcome by local density of a GEO
Does the *repetitive* release of agricultural applications have specific impact?	Agricultural pests (*e.g.*, weeds or insects) may built up resistances to toxins or pesticides
Does 'contained use' imply that *biological containment* need not be considered?	In many applications *absolute* physical containment is not possible
Does a specific GEO application stimulate transboundary movement?	Different geographic regions may have different 'biodiversity' and thereby different susceptibilities
... other relevant questions...?	... include in SRQ...?

FIGURE 5.9: Schematized representation of possibly relevant questions on the *application(s)* of GEO hazard in a procedural and modular format for a catalog of familiarity from which a selection can be made to arrive at a sufficient *set of relevant questions* for the hazard identification of a particular GEO release. An assumed SRQ can be adapted and changed with the precision of individual questions to be included in or excluded from the 'window of concern' for notification. For the present purpose, the *procedure* of choosing relevant research questions is more important than the particular research questions discussed. The right-hand column represents arguments for their relevance (no claim for *completeness* implied).

— 5.7 Setting the research agenda —

The methodological burden of proof for *not* including the cited questions has not been produced by scientific experts *yet.* Therefore, for the time being at least, the methodological status of claims on GEO hazard that are not based on a sufficient addressing of these questions is not scientifically warranted. In practice, this means that the *Competent Authorities* shall have to address these questions or abstain from producing claims on the subject.

This format for a *catalog of familiarity* can be an enabling tool for 'hazard detectives' in their search for relevant research questions. For each particular GEO notification a selection of relevant research questions can be made on the basis of the available *catalog of familiarity.* The result of this selection procedure can be seen as a "flowchart" of questions in which the combination of answers (or lack thereof) will decide wether or not a permit for use of the GEO will be issued to the applicant. For a specific intended release, the questions in the catalog can be 'checked' as being relevant or irrelevant by the *Competent Authority* or its advisory board. Exclusion of a question from concern must be supported by an argument for its *irrelevance* in each specific case. This procedure can provide a methodological basis for distinguishing between "need-to-know" and "nice-to-know" (cf. Zannoni 1995; Bergmans 1995). The difference between the two, comes with a scientific burden of proof: "A scientific discussion of possible effects of genetic modification on the eventual GMO [=GEO] and on its functioning in the environment will yield many questions, but usually only a limited number of those questions are relevant for the identification of environmental hazards of the GMO" (Bergmans 1995: 24).

This does not take away the fact that the specific choice of this "limited number of questions" comes with a scientific burden of proof for including or excluding particular questions. Claims made by a *Competent Authority* or its advisory board will typically carry the implication that they are made on the basis of *sufficient familiarity* without the need to address additional (empirical) research questions. An international initiative to assemble a *catalog of familiarity* would amount to a *practical,* relatively *low-budget,* and scientifically *feasible* project to bring together the world-wide existing expertise which is too dispersed now in many cases. The joint effort could also result in a net reduction of the total investment of time and funding that is now required of the separate advisory boards and *Competent Authorities* around the world.

Generic review will mean in practice that certain questions need not be addressed for some categories of GEOs. Accumulating experience may lead to routine exclusion of *some* questions for *some* kinds of GEOS UNTIL the argument for exclusion or irrelevance should be successfully challenged again. The *Competent*

Authorities are interpreters of the familiarity criterion and can only perform that task by the use of a legitimate scientific basis for their interpretation. Advisory boards such as the Dutch COGEM play an important role in the process of deciding which questions to include in a GEO hazard identification.

Many participants in the debate have expressed concern about a limited availability of relevant knowledge (cf. Regal *et al.* 1986). Bergmans (1995) notes in relation to *microbial ecology* that, "..., only an increased knowledge of the basic natural processes can help risk analysis" (Bergmans 1995: 25).

Thinking and discussing in terms of relevant research questions will not just serve the purpose of biosafety assessment, but it will also be an immediate help in relation to the *agenda-setting* of relevant biological background research. Knowing what it is that we would like to know, will help us develop a research programme that might produce this knowledge. The more specifically the relevant research questions are articulated, the more specifically will we be able to decide whether we can answer them (now or in the future) or not and the closer will we get to a scientific biosafety assessment. The *categories* of questions for familiarity in GEO hazard identification may function in practice as a *heuristic tool* or searching device for uncovering relevant research questions that need to be addressed.

As it is, applicants for GEO permits are too free to deliver "empirical data" of their own experimental design on possible hazards of a proposed GEO application. Maybe it should be considered to make a more clear-cut regulatory separation between permit applicants and hazard researchers. Applicants could be asked to provide research *funding* rather than their own hazard identifications. The SRQ approach may also provide a useful resource for the design of *monitoring* requirements. The questions to be posed in the monitoring process may also be drawn from the developing catalog of familiarity.

Discussing and setting the research agenda is vitally important. Without an adequate methodological analysis of relevant research, the investments made in biosafety assessment are largely wasted. An international group of experts concluded in a study for the OECD: "In biosafety terms, it is clear that the amount of useful information gained from releases to date has been limited (...) The releases have in effect been 'field containment' rather than true releases" (OECD 1993: 23-24; cit. in von Schomberg 1998: 7). Bergmans (1995) has argued that, "[i]n order to find out what the environmental effects of GMOs are, one should (...) plan experiments with an expected environmental effect" (Bergmans 1995: 25). For this purpose too, a catalog of familiarity can be a guide for relevant hazard research.

In practice, any SRQ will have to be limited. If for no other reason, then at least to limit the *bureaucratic burden* for the regulatory apparatus (cf. Barnthouse 1992; Chesson 1990; Levin 1990). On the other hand it is questionable whether more specific information requirements will necessarily lead to 'thicker files' and more time-consuming procedures. More specified relevant questions for notification may also lead to less uncertainty for the applicant or notifier as well as for the overseeing *Competent Authority*. Where the *Competent Authority* chooses to limit an SRQ and exclude possibly relevant questions for which the scientific burden of proof for exclusion has not been produced, the *Competent Authority* will take on a *political* responsibility in those cases.

CHAPTER 6

A Moral Responsibility for Scientific Experts

— 6.1 Choices in biosafety management —

In the introductory chapter of this study, a policy-maker was cited who found himself faced with the problem of having to cope with controversy between "orthodox" and "unorthodox" experts. Now we have the analytical apparatus to distinguish the scientific kernel of a controversy from its political contention. It implies that both "orthodox" *and* "unorthodox" expertise can be interpreted and made productive in a procedural framework of evaluation.

If politicians with a democratic basis of legitimation decide not to address some possibly relevant question(s), then this is a political decision in its proper context. If scientists take such a decision, then this is a political decision in an *improper* context. Whereas science and politics cannot always be *separated* in an applied context such as biosafety assessment, it is a primary concern for democratic societies to ensure that the two are at least *distinguished* in the context of practical decision-making.

Choices in biosafety management must be scientifically informed. References made in the biosafety debate to such terms as 'plausibility', 'familiarity', 'acceptability', 'similarity', 'evidence', etc., are examples of how *science* and *society* cannot be separated in practical decision-making. To safeguard the quality of decision-making, an operational distinction between scientific and political perspectives and their respective bases of legitimation is required. Amalgamating the two leads to what Regal has called, "cryptic philosophy and ideology in the 'science' of risk assessment" (Regal 1996).

Failure to distinguish between science and society in technology assessment will lead to what I have called 'artificial controversy' in applied science (see Section 1.6), and will do more to spread the smoke of a debate than to recognize the fire of it. Paraphrasing the representation of Wittgenstein cited earlier, we may note that 'language is not just the vehicle of a *controversy*, but also the driver' (cf. Section 1.2.1). Ill-defined language will easily take a debate away from the problem at stake.

Different ways have been suggested to analyse the pressing challenge of reconciling science and society in a *technological culture* (cf. Jasanoff *et al.* 1995; Toulmin 1990; Regal 1990). The focus of the present study is on the methodological basis of scientific claims that are contested. Discussion about the relevance of research questions can provide a medium of translation between *Cosmos* and *Polis*, between *science* and *society* (cf. Toulmin 1990). Translations from science to society and *vice versa*, can be made through the medium of relevant questions for understanding practical and theoretical problems. The presented approach allows for a pragmatic interpretion of such thought-bridges as *plausibility*, *familiarity*, *evidence*, *similarity*, *acceptability*, etc.; both from the point of view of science *and* from the point of view of society. Thus, it enables policy-makers and administrators to acknowledge the relevance of "orthodox" as well as of "unorthodox" expertise. Given the complexity of assessing modern technology, we cannot do without the cunning of our *scientific detectives.*

Science researchers have expressed concern about, "How science fails the environment" (Wynne and Mayer 1993). What triggers the perception that sometimes scientific experts seem to be adversaries of *environmental prudence* rather than its supporters? In my analysis, it is not because science does not have the methodological *potential* to produce prudent claims (what could be more prepared for this task than science?), but rather because scientific experts do not always live up to the *responsibility* of taking their own scientific methodology seriously. This is a grave concern because, as Tversky and Kahnemann (1974) have observed, "..., people respond differently when given no evidence and when given worthless evidence" (Tversky and Kahneman 1974: 1125). Scientific expertise remains the best resource for validating and evaluating the relevance of evidence.

6.1.1 Demarcating science from 'transscience'

To be able to distinguish science from society in decision-making, an important prerequisite is to distinguish applied science from *transscience* as defined by Weinberg (1972): "Many of the issues which arise in the course of the interaction between science or technology and society – *e.g.*, the deleterious side effects of technology, or the attempts to deal with social problems through the procedures of science – hang on the answers to questions which can be asked of science and yet *which cannot be answered by science.* I propose the term *trans-scientific* for these questions since, though they are, epistemologically speaking, questions of fact and can be stated in the language of science, they are unanswerable by science; they transcend science" (Weinberg 1972: 210 – no italics added; cf. Weinberg 1985).

Two examples of "transscientific questions" that are mentioned by Weinberg (1972) pertain to, "biological effects of low-level radiation insults" and, "the probability of extremely improbable events" (Weinberg 1972: 210-211). Both examples imply research questions that would take so extensive or long-term testing, that, as a practical matter, they are, "unanswerable by direct scientific investigation" (Weinberg 1972: 210). Traditionally, philosophy of science has focused on demarcating science from *pseudoscience.* An attempt to shed light on the practical chemistry between science and society must also address the demarcation between science and *transscience.* An approach of this distinction from within science and its methodology is required to recognize situations in which a question is considered to be 'transscientific', while in fact it falls under the methodological competence and basis of legitimation of scientists to answer it.

"Transscientific questions" should be answered politically, not scientifically. In this respect I disagree with Van den Daele, who claims that: "Experts are not simply applied scientist; expertise is mostly 'transscience'" (van den Daele 1992: 331). According to Van den Daele, "[t]his gap is bridged through intuition, loose reasoning by analogy and 'educated guesses'" (van den Daele 1992: 331). To amalgamate scientific and political assessments in such an intuitive way, would mean that separate *contexts of relevance* get mixed up. In the process it cannot be avoided that scientists will be tempted to make political claims and politicians will be tempted to make scientific claims. This temptation, which is lurking in the very terms of the biosafety debate, poses a threat to rational and democratic choices in biosafety management. If scientists yield to answering "transscientific" questions, they go beyond their methodological mandate.

It should be noted that many *scientific* questions cannot be answered by science either. This implies that science can only be a resource for *part* of the larger puzzle of policy decisions, albeit an indispensable part. The *demarcation* of scientific and transscientific questions is first and foremost a *responsibility* of scientists, since they are in the best position to judge which questions can (or could) and which questions cannot be answered by science. To recognize the fallacies of interpreting and addressing *transscientific* questions (with normative answers) as if they were *scientific* questions (with empirical answers), and *vice versa,* is a *moral responsibility of scientific experts.*

Given the fact that we cannot have all the *answers,* we may at least strive to have an overview of (most of) the relevant *questions.* The possible 'windows of concern' for biosafety assessment are *defined* by scientists and *selected* by politicians. The latter may choose between alternative possible SRQs, but they cannot legitimately alter the scientific problem definitions that they reflect. Scientific experts have the methodological task to *raise* relevant research questions, even if these questions cannot (yet) be answered by science. This implies that without science, we cannot recognize the relevant *transscientific* issues either.

An example of how *artificial controversy* may lead scientists to answer *transscientific* questions is the recurring claim that genetic engineering is a necessary prerequisite to be able to *feed the world population* in the future. Through analysis of the artificial controversy, we may arrive at uncovering the scientific questions involved in this claim and to demarcate them from the transscientific questions. Another example of a 'transscientific' question is the question whether or not we are sufficiently 'familiar' with a GEO release to identify its possible hazards (see Chapter 5). As demonstrated in the previous chapter, this question can and must be made *scientific* by listing the relevant research questions that need to be addressed before it can be dealt with as a *transscientific* question in a policy or administrative context. For a scientific expert, *as a scientist,* to answer a transscientific question would be a methodological trespassing of the legitimate boundaries of scientific research.

'Science' and 'transscience' are both about asking questions. To demarcate the two, a distinction must be made between the *types* of questions that the two notions imply. Scientific questions can be evaluated in terms of their *methodological relevance,* transscientific questions should be evaluated in terms of their *political relevance.* Important for their demarcation is that 'artificial controversies' are recognized for what they are. This can be done by taking special care for the way the questions are raised and addressed. The type of answers that can be given to *transscientific* questions as suggested by Van den Daele (1992) belong to the realm of *politicial* legitimation rather than to the realm of *scientific* justification. Van den Daele fails to specify this distinction. Failure to distinguish 'artificial' from 'fundamental' controversies, may also tempt scientific experts to give *sientific* answers to *transscientific* questions. This is a *category mistake* in the sence that *empirical* answers are given to *normative* questions. This is another manifestation of the classical *naturalistic fallacy.* Avoidance of this normative pitfall can be considered as a *moral responsibility of scientific experts.*

One ecologist raised the rhetorical question in relation to the *political* relevance of GEO biosafety assessment: "So what?" (Crawley 1994: 48). It should be kept in mind that some simply deny the existence of controversy over biosafety issues (cf. Miller 1997). His own answer to this question is: "The reason that governments treat GMOs [=GEOs] differently from other kinds of introduced organisms (like biocontrol agents or new garden plants) is because of fear and uncertainty. Most of the fears are completely unjustified, but not all (...) The worst way to deal with fear and uncertainty is by denial (*e.g.*, by saying that

GMOs are just like any other organisms, so there is nothing to be concerned about). This is an example of what we might call the *'Trust me. I'm a doctor'* syndrome" (Crawley 1994: 48).

The element of 'trust' can only be preserved by applying the highest standards of quality control to the products of science (cf. Sloep 1993). One prerequisite to safeguard these standards is to demarcate scientific from transscientific questions, so as to avoid situations in which scientific experts give 'scientific' answers to 'transscientific' questions.

— 6.2 *Scientific* questions and *political* answers —

Political considerations cannot replace scientific considerations. The same is true for ethical considerations. We can decide to let ethical concerns prevail over scientific concerns, but they cannot replace them. In practical ethical questions, scientific questions will play a role also. Depending on the scientific questions addressed, the answer to ethical questions may even change. This makes it important to consider scientific questions as indispensable ingredients of political, ethical and other normative contexts of reasoning.

According to Wöhrmann *et al.* (1996), when it comes to dealing with biosafety issues, "[t]he final decision will always be a political and ethical one" (Wöhrmann *et al.* 1996: 1). I agree with this general assessment, but I also believe that on the way to such final decisions science plays a 'decisive' role. For example, a Dutch committee on the ethical aspects of agricultural uses of biotechnology has suggested an ethical imperative for applying genetically engineered plants in terms of a "no, unless" directive: genetic engineering of plants will *not* be allowed, *unless* there are practical reasons to overrule the prohibition (cf. Kockelkoren 1993). As I have argued elsewhere (van Dommelen 1996b), the "unless" part of this directive represents an *empirical evaluation* rather than an *ethical evaluation.*

The relationship between science and ethics has received much attention in the *social studies of science* (cf. Longino 1990). The *methodological analysis* of the present study opens up a perspective on this relationship also. The choice between one or another *set of relevant questions* (SRQ) as a methodological basis for scientific research may have important *moral implications* for scientific expertise. If it turns out that one assumed SRQ cannot support some specific scientific claim, then this claim would be scientifically unsubstantiated. This puts a *moral* responsibility on those who act as scientific experts to critically review the methodological status of their scientific claims, especially since this review cannot be done by those who lack the required scientific literacy. Political or ethical evaluations cannot be made without a clear conception and grasp of the *methodological status* of scientific claims.

One concern in the relationship between values and science in which the responsibility of scientific experts should be at centre stage is the problem of *scientific bias.* In a crude form one may think of the wanton and strategic use of simplified or misguided arguments by scientific experts. The high political stakes and industrial interests surrounding the development of genetic engineering are certainly an 'excellent' context for the strategic use of a scientific guise for political claims on biosafety. A scientific expert, *qua scientist,* has a responsibility to safeguard the quality and relevance of empirical data. Thus, a concern such as expressed by one of the Dutch NGOs in relation to a GEO notification should be a concern for scientific experts as well: "The safety of herbicide-resistant crops is proven [in this case] by referring to internally conducted, unpublished studies by company XXX, which has invested millions of dollars. This does not create transparency and a climate of trust from the perspective of non-governmental organizations" (NSNE 1997: 92). This type of vested interests poses a serious concern to the unbiased analysis and evaluation of biosafety controversies.

It would be a mistake, however, to completely rely on *social-economic* and *psychological* explanations for the understanding of scientific controversies whenever we come across some, "patently absurd line of reasoning", which could be interpreted as a case of strategic use of scientific expertise (Kareiva and Parker 1994: 8). A one-sided *social-constructivist* understanding of scientific controversies may lead us to overlook the internal *methodological* pitfalls of scientific expertise. For the purpose of specifically studying the latter concern, I take the freedom to presuppose the 'good faith' of all scientific experts involved in biosafety controversies. This approach may help to uncover the fact that even the highest moral standards could not, by themselves, provide a guarantee that the methodological restrictions of scientific inference will be observed. A methodological analysis of biosafety controversies may show how this restrictedness of applied science as a basis for political choices, has a moral status in its own right.

Some have argued that *general value systems* should be seen as the key to understanding scientific controversies. In his analysis of biosafety controversies, Szanto (1993) has argued: "Once the value background has been discovered, we shall perhaps see the history of both the controversy and the recombinant DNA technology in a somewhat different light" (Szanto 1993: 243). Considering the basis of scientific controversy more generally, Szanto concludes: "Thus *values* are essential factors in controversies, and together with interests, needs, and beliefs ("knowledge"), they constitute worldviews. The opposition and struggle of worldviews can be regarded as *the essence of controversies.* (...) Epistemic factors are secondary in this respect" (Szanto 1993: 259 – no italics added).

Are epistemic and methodological factors secondary when it comes to interpreting controversies? This would imply that contested science is a lesser concern for our general understanding of biosafety controversies. But is it? Szanto considers "world views" to be the key to understanding controversy in applied science. It is interesting to note that in the context of biosafety controversies, Regal (1996) sees it as a *distorting problem* that world views rather than science sometimes dominate the scene when it comes to GEO hazard identification. He raises the question: "How can non-experts claim to be making 'scientific judgements' if they do not gather scientific data and are not ready to detail in print how they evaluate what data there is? Their defense is based primarily on world view (or Weltanschauung) that is not recognized to be world view" (Regal 1996: 16). Answering this question, he cites examples of what he calls: "... ancient philosophy and modern ideology that have been passing as empirical science" (Regal 1996: 16). His conclusion is that: "... a great deal of risk assessment in biotechnology has been philosophical beliefs that do not recognize themselves to be philosophy" (Regal 1996: 22). Thus, Regal sees the implicit role of world views in biosafety assessment as a source of distortion rather than as a sound basis for decision-making.

Collingridge (1980) has dedicated a specific analysis to the general problem of the relationhip between *values* and *science* in applied contexts. Considering the relative importance of values and facts in the context of risk assessment, he comes to the conclusion that: "Although any debate about monitoring is an evaluative one – it is about what ought to be done, what is the best option – *the main elements in the debate are always issues of fact*" (Collingridge 1980: 169 – italics added). He arrives at this conclusion by considering the logical structure of evaluative arguments (c.f. van der Steen 1993). This implies that the *normative* conclusion may be affected by raising (new) *empirical* questions. I apply this analysis to reconstruct the role of scientific claims in normative evaluations. Apart from any ethical position, the science involved will also matter for the outcome of policy choices.

This implies that we should not conclude on faulty grounds that some question(s) should be labelled 'transscientific' before it is clear that they cannot be further analysed as 'scientific' questions.

6.2.1 Empirical premisses and normative conclusions

Consider a reconstruction of the possibility to use *ethical evaluations* as a basis for *technology assessment* (cf. Verhoog 1993). In the German TA study of herbicide-resistant plants that was reviewed in Chapter 3, attention was given to ethical evaluations also. According to Altner, who was one of the experts involved in the herbicide-resistance TA process, from the point of view of, "einer biozentrischen Ethik, die Respekt vor der Natur 'um ihrer selbst willen' fordert"[29] (van den Daele *et al.* 1996: 246), there is a *moral imperative* to guarantee, "an acceptable harmony between the goals of nature and human benefits" (van den Daele *et al.* 1996: 247). One of the conditions specified by Altner (1996) that would affect this moral guarantee is when the HR technology, "puts a burden on the physiological stability of the involved plant" (van den Daele *et al.* 1996: 247). Here again, we see that a *normative* conclusion is dependent of the specific content of an *empirical* claim.

Reconstructing the normative and empirical elements of this ethical argument as a syllogism, we get:

Premiss 1. 'If the physiological stability of a plant is burdened by genetic engineering, this will be a morally inacceptable disruption of the harmony between nature and human benefits.'

Premiss 2. 'The physiological stability of a plant is burdened by genetic engineering.'

Conclusion. 'Genetic engineering of plants will be a morally inacceptable disruption of the harmony between nature and human benefits.'

Depending on the methodological support for the empirical claim (*Premiss 2*), the normative conclusion will change face. If scientists cannot argue that the physiological stability of a plant will be burdened by genetic engineering, then the ethical conclusion will be different. If, on the other hand, scientists can argue for the physiological loss of stability in transgenic plants, then this would indeed lead to an ethical embargo given the ethical norm (*Premiss 1*). The coordinators of the German TA study are quite decisive about the empirical claim (*Premiss 2*): "Offenbar stört die übertragung von HR-Genen die physiologische Stabilität der Pflanze nicht nennenswert"[30] (van den Daele *et al.* 1996: 247), but this may also be seen as an attempt to see consensus where other scientific ex-

29 "A biocentric ethics, which demands respect for nature 'for it's own sake'".

30 "Apparently, the transfer of HR-genes does not disturb the physiological stability of plants considerably".

perts see controversy (cf. Weber 1996). The example shows how scientific questions can be decisive for the outcome of ethical answers. Even in ethical decisions sufficient attention must be given to scientific considerations.

Consider another recent episode in the biosafety debate in which science and ethics have been amalgamated. It is representative of the way in which biosafety controversies may touch upon the very roots and sinews of our culture. According to Potthast (1996), "... positions concerning 'evolution' necessarily consist of a combination of arguments from science, philosophy of science and normative ethics" (Potthast 1996: 228). To avoid unproductive artificial controversy, opposing positions must be specified in terms of the research questions that are considered relevant.

Where Potthast (1996) stresses the contrast between views of nature as "dynamic versus stable" (Potthast 1996: 232), I would argue that this is not a fundamental controversy because it is not a matter of *either-or* but rather of *and-and.* Ecosystems will sometimes (or in some respect) be more *dynamic* and sometimes (or in some respect) be more *stable.* It would be a strong claim (with a 'heavy' burden of scientific proof) to hold that any ecosystem could be described or understood *without* reference to aspects of either its dynamics or its stability. In final analysis, the kernel of the discussion is to specify what one considers to be a sufficient SRQ for a specific research purpose. These questions will be concerned *both* with ecosystem dynamics *and* with ecosystem stability.

A pragmatic SRQ approach also sheds light on the legitimation of *biosafety regulation.* As Potthast (1996) argues, "... there have been different 'ways of substantiating' the claim that evolutionary processes are worth protecting..." (Potthast 1996: 228). As *categories of substantion* for this claim, Potthast mentions: "First, aspects of *usability* concerning raw materials, breeding and pharmaceutical chemistry; second, *aesthetics* and *moral value* (here, the claim to satisfaction of *scientific curiosity* is often mentioned as well); and third, an *inherent value* of the processes of evolution beyond all human interests (but humans have to recognize, accept, and operationalize this inherent value in the context of their activities)" (Potthast 1996: 229 – italics added).

Although concerns such as "aesthetics" and "inherent value" certainly have relevance in a *political* arena of debate, they cannot provide legitimation in the *scientific* arena of debate. From a scientific point of view, the issue of GEO hazard identification is whether the release of a GEO may have an impact on existing biological systems. Whether this impact is *acceptable* or maybe even *desirable* is a political issue that may be answered differently in different times and places. It would be an example of not sufficiently demarcating *science* from *transscience* to amalgamate the scientific assessment of an environmental impact with the political assessment of its (un)desirability. As noted earlier, modern agriculture can be described in the metaphors of *warfare* – the acceptability of this practice is not for scientists to decide.

Potthast (1996) omits to mention the substantiation of protection that is the most relevant for biosafety assessment, the imperative of *sustainability*. The fact that humankind has had an impact on biological evolution over centuries, long before the dawn of genetic engineering, does not imply that *new* ways of exerting environmental impacts should not be mistrusted for its consequences. One of the *moral* imperatives involved with the notion of 'sustainability' is that one should not undertake activities about which there is *scientific* uncertainty about their impact (cf. Kasanmoentalib 1996). This (moral) responsibility involved with the notion of 'sustainability' puts a focus on the way one will deal with (scientific) knowledge (cf. Hueting and Reijnders 1998). The normative concern with knowledge of sustainability has become part of the policitical debate about biosafety in the form of the *precautionary principle*, which is essentially an imperative in relation to the *methodological basis of scientific knowledge*. Precautionary science is aware of the limitations of its claims and attempts to recognize an existing lack of information with the help of *critical scientific methodology* – which should be the basis of all scientific research.

— 6.3 Precautionary science for a sustainable future —

Genetic engineering has been promoted as a technological tool to arrive at *sustainable development*. A group of specialists appointed by the *Council of Europe* examined the "Ecological Impacts of Gene Technology", and concluded that: "Biotechnology can contribute to sustainable development" (van der Meer 1994: 11).

An associated policy tool for decision-makers to cope with scientific uncertainty is the so-called *precautionary principle*. At the *Earth Summit* of the United Nations in Rio de Janeiro in 1992, from which *Agenda* 21 resulted, it was agreed in "Principle 15" of the *Rio Declaration on Environment and Development* that: "In order to protect the environment, the precautionary approach shall be widely applied by States according to their capabilities" (UNCED 1992: 10). This encompassing approach for dealing with the scientific basis of environmental decisions was phrased as follows: "Where there are threats of serious or irreversible damage, lack of full scientific certainty shall not be used as a reason for postponing cost-effective measures to prevent environmental degradation" (UNCED 1992: 10). The precautionary principle can be reconstructed along the same lines of argumentation as normative inferences more generally can be reconstructed.

O'Riordan and Jordan (1995) say about the political potential of the *precautionary principle:* "It's future looks promising but is not assured" (O'Riordan and Jordan 1995: 191). One of the primary prerequisites for giving the precautionary principle a fair chance to define our future is a proper analysis of the role and methodological status of scientific controversy. Giving due attention to the *limits* and *restrictions* of scientific methodology is a moral responsibility for scientists involved in applied expertise. Without responsible scientific expertise, it is not possible to make *precautionary technology assessments.*

The precautionary principle has been presented as a challenge to, "established scientific method" (O'Riordan and Jordan 1995: 193). According to O'Riordan and Jordan, "the problem for science in the precautionary mode is that its normal reliance on experimentation, theory falsification, verification, consistency and predictability is thoroughly challenged" (O'Riordan and Jordan 1995: 199). In contrast to this view, I would argue that the precautionary principle represents an attempt to *take science seriously,* not a method of discarding or revising it. A proper application of the precautionary principle must be firmly based on adequate scientific methodology.

Although the precautionary principle has made an impressive career since its inception in Germany in the 1970s, there is still confusion about its interpretation: "while it is applauded as a 'good thing', no one is quite sure about what it really means or how it might be implemented" (O'Riordan and Jordan 1995: 194). A crucial step towards clarifying this confusion is the development of a practical interpretation of scientific controversy. Without this methodological prerequisite there can hardly be hope to find a proper way to apply the *precautionary principle* and to interpret its normative evaluations.

As was shown in previous chapters, controversy prevails in the context of biosafety assessment. Can this be considered as a "lack of full scientific certainty" and thus as a sufficient basis to invoke the precautionary principle? The answer to this question will depend on the underlying view of what constitutes a scientific controversy. If science plays a fundamental role in policy, regulatory and ethical decisions, then what are the implications of scientific *controversy* for the decision-making processes? Specifying the relevant research questions is a prerequisite for using the precautionary principle in an applied context.

Presented in the form of a syllogism, we can reconstruct the normative and empirical aspects of the *precautionary principle* as follows:

Premiss 1. If there are threats of serious or irreversible damage, lack of full scientific certainty shall not be used as a reason for postponing cost-effective measures to prevent environmental degradation. (= precautionary principle)

Premiss 2. There are threats of serious or irreversible damage. (= contested scientific claim)

> *Conclusion.* Lack of full scientific certainty (about the supposed threats) shall not be used as a reason for postponing cost-effective measures to prevent environmental degradation.

Premiss 1 and the Conclusion are both normative statements. *Premiss 2*, however, is an empirical statement. In this elementary logical reconstruction of the precautionary principle it becomes clear that we must still find ways of evaluating the contested empirical claim of *Premiss* 2, before we can decide whether the normative *Conclusion* ought to be accepted (given the acceptability of the normative *Premiss 1*). This shows us that a precautionary approach to environmental questions cannot be implemented without a proper analysis of scientific controversy as discussion about sufficient SRQs for the research purposes. If *Premiss 2* is contested, this will change our perspective on what it means to "proceed with caution", and this may eventually result in the decision not to apply the precautionary principle.

Applying Collingridge's analysis (see Section 6.2) to the precautionary principle, we see that it is science which initially triggers precaution (cf. Ruesink *et al.* 1995: 471). To use the precautionary principle in practice one must still rely on the scientific community for information about the scientific state of affairs. If *scientific* experts claim that there is *no*, "lack of full scientific certainty" in relation to a specific environmental problem, then there is no *political* basis for invoking a precautionary approach. Therefore, to arrive at a proper implementation of the precautionary principle, one cannot get around producing a proper interpretation of what constitutes relevant scientific knowledge in a particular context.

How can scientific evidence be evaluated when it is contested? Only in the case of sufficient *scientific* reason for concern, does the precautionary principle imply that there is sufficient *political* reason for a precautionary approach. The kernel of the precautionary principle is its focus on the burden of proof for including or excluding research questions as relevant or irrelevant to our *window of concern.*

6.3.1 Missing relevant questions and research bias

As an example of how a responsible interpretation of scientific methodology may enhance precautionary biosafety management, consider the following. It is an illustration of the pervading methodological importance of thinking about the relevance of research questions for specific research purposes and an important concern in the context of GEO hazard identification also. In practical situations, not addressing a question adequately is equivalent to not addressing that question at all. The way in which a question is addressed is therefore an important responsibility of scientific experts, with moral implications.

The *methodological relevance* of empirical data that are cited to support a specific claim depends upon the specific empirical research questions on the basis of which those data were collected. Consider the way experimental data are evaluated in *statistical analysis.* Statisticians distinguish 'Type-I' and 'Type-II' errors as possible methodological fallacies in the interpretion of a data set. Peterman and M'Gonigle (1992) argue that Type-I errors usually get the most attention, but Type-II errors are no less interesting in the context of a precautionary hazard identification (cf. Buhl-Mortensen 1996; Peterman and M'Gonigle 1992). Type-II errors can occur when a data set has low *statistical power* in relation to the research question (typically expressed in the form of a *test of alternative hypotheses*, H_0 and H_1). A data set which is ill-adapted to a specific research purpose can give rise to a misguided conclusion of "no observed effect" on the basis of the experiment from which the data were drawn, while in reality the hypothesized effect *does* occur. Statistical power is defined as $1-\beta$, where β is the *probability of failing to reject the null hypothesis of "no effect" (H_0) when in fact H_0 is false.* Statistical power thus is a measure for the probability of correctly rejecting H_0 (Peterman 1990: 4; see figure below).

	DECISION	
STATE OF NATURE	*Do not reject null hypothesis*	*Reject null hypothesis*
Null hypothesis actually true	1. Correct ($1-\alpha$)	2. Type I error (α)
Null hypothesis actually false	3. Type II error (β)	4. Correct ($1-\beta$) (= power)

Whenever data are collected on the basis of a set of assumed relevant empirical questions in which a specific biological mechanism or *effect* (see Sections 5.6.1-5.6.7) has not been sufficiently included, then the statistical power of those data is likely to be low for detecting that mechanism or effect (cf. Peterman and M'Gonigle 1992: 232). Thus, Type-II errors of statistical interpretation must be recognized and avoided by giving due attention to the relevant empirical questions for a specific research purpose. When data are collected on the basis of a less than sufficient set of questions, the statistical power of a data set will decrease. The relevance of a data set for testing a specific hypothesis always depends on the way the data set was generated. Since different empirical questions will produce different data sets, the latter cannot be put forward in a debate as "evidence" without reference to the assumed relevance of the questions. If statistical power of a data set for a specific impact is low, then, "the monitoring programme may not be a reliable source of information because it is likely (with probability = 1 - power) to have failed to detect such an effect, even if it was present. In that case, little confidence should be placed in results from the monitoring programme" (Peterman and M'Gonigle 1992: 232).

Peterman and M'Gonigle (1992) give a methodological warning that, "statements about safety or lack of an important effect are associated with an agreed level of statistical power" (Peterman and M'Gonigle 1992: 232). Peterman (1990) found that this warning is not as *redundant* in scientific research as one might expect it to be. In a survey of papers in the *fisheries and aquatic sciences* he found that no less than 98% of empirical studies that did not reject some null hypothesis failed to report either the probability of making a type II error (1-β), or statistical power (β) (Peterman 1990: 2). A proper interpretation of experimental results can only be done by explicit reference to the questions that were used as a basis for the experiment, specified as the assumed relevant questions for a particular research purpose.

In the context of GEO hazard identification, this problem has been studied by Köhler and Braun (1995). In their view, the purpose of GEO hazard identification is especially sensitive to concerns about Type-II errors in statistical data interpretation: "Schwerwiegender in bezug auf eine Freisetzung von transgenen Nutzpflanzen kann aber der β-Fehler sein. Er gibt die Wahrscheinlichkeit an, mit der die Nullhypothese H_0 beibehalten wird, obwohl die Alternativhypothese H_1 gültig ist. In diesem Falle (...) würde man irrtümlicher Weise annehmen, daß die Häufigkeit der Transgene geringer ist als der Schwellenwert S. Man verzichtet dann auf entsprechende Maßnahmen, obwohl sie angebracht wären. Der β-Fehler beinhaltet daher im wesentlichen das Ökologischen Risiko ..."[31] (Köhler and Braun 1995: 178).

Possible *methodological pitfalls* such as these can only be recognized and avoided by the scientific experts who are involved with practical problems such as biosafety assessment. Only scientists can provide the required safeguarding from these pitfalls. Without their qualified input, those responsible run the risk of dealing with *scientific* questions as if they were *transscientific.*

---

31 "More serious in relation to a release of transgenic plants can be the β-error. It indicates the probability with which the null hypothesis H_0 is sustained, although the alternative hypothesis H_1 is valid. In this situation (...) one would mistakenly assume that the abundance of transgenes is less than the threshold value S. In such a case one would abstain from adequate measures, although they would be appropriate. The β-error thus essentially represents the ecological risk ...".

— 6.4 Science as the gatekeeper of 'not-knowing' —

The history and methodology of science is typically associated with the 'denial' and 'exposure' of unjustified or supposed knowledge such as *revelation* or *intuition.* Science can be seen as the art of recognizing the *methodological conditions for reliable knowledge.* As such, the sciences are a wonderful cultural asset that is an irreplacable resource for *sanity* and *democracy.* Von Weizsäcker has addressed concern about biosafety assessment as a dilemma between, "Lacking Scientific Knowledge or Lacking the Wisdom and Culture of Not-Knowing?" (von Weizsäcker 1996). The two are both historically *and* methodologically intimately related. As Van den Daele (1994) remarked in relation to the technology assessment of genetically engineered herbicide-resistance: "Letztlich geht es um die Entscheidung wie wir mit Nichtwissen umgehen sollen"[32] (van den Daele *et al.* in: Sukopp and Sukopp 1994: 141).

From the reconstructions presented in Sections 6.2 and 6.3, it becomes clear that there is no way around *scientific controversy* when it comes to *biosafety assessment* and *regulatory policy.* Both in applications of the precautionary principle and in general normative evaluations, we still need to deal with contested science and must have a practical analysis of scientific controversy. Different suggestions have been made about how, "to learn to understand what science is" (Hey 1996: 97). The SRQ approach in combination with the methodological burden of proof as a selective tool for relevance, gives policy makers *and* lay audiences a better grasp on contested expertise.

The main concern for controversial science as a basis of biosafety assessment is to think about what constitutes a sufficient SRQ for the research purpose. This approach gives a practical key for dealing with scientific uncertainty. It may help to make the existing uncertainty more conceivable and to bring it out into the open. If it makes sense to see scientific research as the *art of asking questions,* then it may also make sense to see a *moral responsibility* for scientists to give utmost care to recognizing and addressing possibly relevant questions for the purpose of GEO hazard identification.

Clarifying how to deal with scientific *controversy* may also help to put the responsibility for acting under the circumstances of scientific *uncertainty* or *ignorance* where it belongs, in the hands of democratically chosen politicians. An important function of scientists in this process is to uncover those uncertainties, not to provide reassuring assessments. In the context of biosafety assessment, an important task of science is to clarify what it is *we do not know* (cf. von Weiz-

32 "In final analysis, the decision of how to deal with not-knowing is at stake".

säcker 1996), to uncover the known unknowns. One of the main cultural tasks of science is to protect us from thinking that we know more than we actually do.

6.4.1 Public perception and 'error-friendliness'

Complaints about "ignorant lay audiences" and "problematic publics" (Davison *et al.* 1997) should be a concern of scientific experts also. Scientists have a responsibility for clarifying what is *not known*. The proper way to do this is to produce relevant research questions and thereby clarify what information is needed and perhaps lacking. This implies that adequate interpretation of *opinion polls* (for example, the *Eurobarometer*), requires sufficient insight in the 'image' of scientific knowledge and its reliability among the general public. Since public perception is an important basis for political legitimation, it is an important concern for biosafety management (cf. Hill and Michael 1998). At the same time, public perception of biosafety issues cannot be seperated from a society's scientific literacy.

Recognizing uncertainties in scientific knowledge by methodological analysis of controversial empirical claims, may add to the awareness that the potency of science is limited. The limits of science are a good reason for policy makers to opt for practical applications with an appropriate "error-friendliness", considering the scientific uncertainties of the potential high-impact technology of environmentally releasing genetically engineered organisms (cf. von Weizsäcker 1995, 1996; cf. Weber 1996: 23). Whereas the *precautionary principle* deals with the availability of the relevant *science*, the notion of *error-friendliness* deals with the possible impact of *technology*. This distinction has the practical implication that the latter notion will be less abstract to a larger public than the notion of 'precautionary science'. From the perspective of biosafety management, the two notions of *precautionary science* and *error-friendly technology* can be considered as flip sides of one coin. Public acceptance will predominantly depend on a technology's image in terms of 'error-friendliness'.

As an illustration, the so-called *millennium problem* (cf. Section 1.1), may show us how the *error-friendliness* of technologies can be a grave concern. If the 'error' of the faulty year-count does indeed lead to trouble at the turn of the century, then we may witness the *widespread dependency* of our society and culture on the reliability of computer technology. At least one interesting parallel between the error-friendliness of *computer technology* and the error-friendliness of *agricultural biotechnology* may be found in the strong *network* relationships of both the *world wide web of computers* and the *global ecological web*, opening possibilities for unforeseen *cascade* effects. The 'appeal' of the notion of 'error-friendliness' as a criterion for the evaluation of the use of a technology may be illustrated by the fact that most probably only relatively few people will be prepared to embark on an airplane flight during New Year's eve of 1999, considering the hazard of malfunction of computers or embedded software.

Applying the imperative of 'error-friendliness' in the context of biosafety management would urge one, for example, to abstain from such practices as mixing genetically engineered soy beans with 'conventional' soy beans (cf. HRH the Prince of Wales 1998). If *public participation* means anything, the public should at least have a choice of *taking* a risk, or not (cf. Rehmann-Sutter 1996, 1998). This also implies that biosafety policy should take into account the importance of the *labelling* of GEO products. Taking seriously the scientific burden of proof for biosafety claims is a prerequisite for making 'error-friendliness' a part of biosafety management. It may be brought closer to home by trying to recover the relevant questions and by rethinking adequate problem definitions for sustainable development.

For the practical purpose of biosafety management, the *categories of questions* that could be relevant for hazard identification as introduced in Chapter 5, may be supplemented and extended by the inclusion of categories such as 'social cost', 'economic cost', 'social benefit', 'economic benefit', 'imposed risk', 'taken risk', 'scale of application', 'alternative technologies to deal with the problem', 'ethical considerations'. Here too, *sets of relevant questions* can be made the focus of debate. The question *who* will benefit in relation to *who* will bear the risk, for example, is an important concern for biosafety policy. The SRQ approach allows for constructive incorporation of scientific expertise from different disciplines. On a basis of complementing SRQs, addressing the scientific *and* political relevance of concerns would become possible. Calls for a - *moratorium* on applications of genetic engineering can thus be interpreted as an appeal to stop and make an inventory of relevant questions before proceeding (cf. Goodwin 1996; Independent Group 1996).

The notion of 'error-friendliness' may be operationalized and related to 'precautionary science' in terms of the relevant questions of concern. It may help *public perception* as well as *scientific expertise* in the process of developing a 'window of concern' that is dedicated to limiting costs and preserving benefits of conventional agriculture *and* agricultural biotechnology. 'Error-friendliness' has always been a concern of sustainable cultures, before "monocultures" (cf. Shiva 1995) promised to be more productive.

— 6.5 Epilogue: Processes of learning —

Unovering the possible future impacts of new technologies is in many respects a *learning process.* One way to learn about potential effects is to be surprised by them. In some cases, the 'surprise' will mean that it is too late to prevent inadvertent consequences. This implies that there will be a *reward* for proper anticipation of potential impacts and a *penalty* for failing to identify possible hazards. Both rewards and penalties may contribute to processes of learning, but the preference will usually not be with learning the hard way.

If it would be clear beforehand which are the relevant questions to be considered for an adequate identification of possible effects, then this would give us an ideal basis for taking decisions about the wise application of new technologies such as genetic engineering. Since this requirement is not realized in many cases, the next best thing to do is to strive for a timely overview of those questions that are possibly relevant for an adequate impact assessment. This process is complicated by all the obstacles that are generally part of the desire to learn something new.

At the same time it should be noted that clarification of the biosafety issues will not resolve all dilemmas in the larger biotechnology debate, which is riddled with ideological, ethical, and other normative evaluations. Scientists can play a facilitating role in these issues by presenting the involved scientific knowledge in a realistic and reliable form. However, as history keeps teaching us, *ideology* and *world view* will not easily be influenced by the results of scientific research. Below, an overview is given of how the analysis presented in this study may be productive in the learning process of technology assessment.

Summing up, the main objectives of this study as listed in Section 1.1.2 are met to the following extent:

- A framework for the *methodological analysis of controversy in science for policy* as a basis for evaluating conflicting scientific claims has been developed in the form of a focus on the fundamental relationship between a specific practical problem and the choice of research questions that are considered relevant for its investigation,
- The developed analytical framework can be applied for the *evaluation of ongoing controversies over biosafety assessment of genetic engineering* by listing the research questions that are considered relevant for the purpose of hazard identification by the contesting parties and thereby making it possible to recognize *artificial controversies* that result from unclarity about the relevance of research questions as an unproductive burden to the debate,
- The developed framework is *fundamental* in the sense that it goes to the core of the involved scientific research by focusing on the methodological basis of

scientific claims in terms of the relevance of the chosen research question given a specified purpose of investigation and the associated burden of proof for defending or denying that relevance,

- The developed framework is *accessible* to many parties in the sense that it can be understood and applied from different perspectives since raising and defending the possible relevance of specific research questions is not privileged to those experts who have access to producing possible answers, but is open to all those who feel concerned about the subject,
- The developed framework is *pragmatic* in the sense that it gives a constructive tool to participants in the debate by providing the communicative format for a cooperative rather than an adversary approach to the development of a sufficient set of relevant questions, thereby making the identification of possible hazards a constructive learning process,
- The developed analytical framework has been demonstrated to make a *significant difference to past, present and future approaches of controversy in science for policy* by applying it to *processes of technology assessment* (Chapter 3), to *conflicting biosafety claims* (Chapter 4), to *policy initiatives for overcoming controversies* (Chapter 5), and to the *interpretation of scientific responsibility* (Chapter 6).

To realize a society that is in *sustainable development* and that can maintain a balance between economy and environment, a sensitive apparatus of permanent learning must be put in place. The balanced development of a dynamic culture will always require feedbacks and precaution. Thinking about the collective learning process of sustainable development in terms of the relevant questions that need to be addressed in different times and contexts, may contribute to maintaining a balance that shows respect for future lives. This concern applies to the development of genetic engineering as well as to other defining components of our technological culture.

The presented analysis entails a number of starting-points and possibilities that may enhance learning processes. These possibilities range from a more detailed concern for the general methodology of applied science, to a more constructive discussion about possibly relevant research questions, to a wider participation of those concerned in the assessment of new technologies, to more transparent decisions of policy-makers and politicians to give green light to the application of new technological possibilities – or not.

The SRQ approach allows for a *dynamic* and *modular* elaboration of the sensitive and sensible 'windows of concern' that are required to open our view to a sustainable future. One difference that the SRQ approach can make in practical situations is that it opens up the potential for making technology assessments more transparant to the involved parties and thereby it may enhance precautionary policies. Processes of learning depend on the possibilities for people to raise their concerns and to ask questions.

References

ABRAC [Agricultural Biotechnology Research Advisory Committee] (1995a). *Performance Standards for Safely Conducting Research with Genetically Modified Fish and Shellfish – Part I. Introduction and Supporting Text for Flowcharts*, Washington, DC: USDA (Office of Agricultural Biotechnology).

ABRAC [Agricultural Biotechnology Research Advisory Committee] (1995b). *Performance Standards for Safely Conducting Research with Genetically Modified Fish and Shellfish – Part II. Flowcharts and Accompanying Worksheets*, Washington, DC: USDA (Office of Agricultural Biotechnology).

Adam, K.D. and Köhler, W.H. (1996). "Evolutionary Genetic Considerations on the Goals and Risks in Releasing Transgenic Crops", in: J. Tomiuk, A. Sentker and K. Wöhrmann (eds.), *Transgenic Organisms – Biological and Social Implications*, Birkhäuser-Verlag, 59-79.

Adelberg, E.A. (1985). "Summary of Proceedings", in: H.O. Halvorson, D. Pramer and M. Rogul (eds.), *Engineered Organisms in the Environment: Scientific Issues*, Washington, DC: American Society for Microbiology, 233-235.

Alexander, M. (1985). "Ecological Consequences: Reducing the Uncertainties", *Issues in Science and Technology* 1 (3), 57-68.

Alexander, M. (1990). "Potential Impact on Community Function", in: J.J. Marois and G. Bruening (eds.), *Risk Assessment in Agricultural Biotechnology: Proceedings of the International Conference (Davis, California, 1988)*, Oakland: Division of Agricultural and Natural Resources, University of California, 121-125.

Altmann, M. (1992a). "'Biopesticides' Turning into New Pests?", *Trends in Ecology and Evolution* 7 (2), 65.

Altmann, M. (1992b). "Reply to J.Th. McClintock and R.D. Sjoblad", *Trends in Ecology and Evolution* 7 (10), 352.

Altmann, M. (1994). "Schlußfolgerungen aus dem Technikfolgenabschätzungs-Verfahren (TA) zum Anbau von Kulturpflanzen mit gentechnisch erzeugter Herbizidresistenz", *GAIA* 3 (6), 309-311.

Altmann, M. (1995). "Transgene Pflanzen und Biodiversität", in: S. Albrecht and V. Beusmann (eds.), *Ökologie transgener Nutzpflanzen*, Frankfurt/New York: Campus Verlag, 127-142.

Altmann, M. and Ammann, K. (1992). "Gentechnologie im gesellschaftlichen Spannungsfeld: Züchtung transgener Kulturpflanzen", *GAIA* 1 (4), 204-213.

Altner, G. (1996). "Ethische Aspekte der gentechnischen Veränderung von Pflanzen (Kurzfassung des Gutachtens)", in: Van den Daele, W. *et al.*, *Grüne Gentechnik im Widerstreit. Modell einer partizipativen Technikfolgenabschätzung zum Einsatz transgener herbizidresistenter Pflanzen*, Weinheim, BRD: VCH, 235-239.

Arts, B. (1998). *The Political Influence of Global NGOs. Case studies on the climate and biodiversity conventions*, Utrecht, the Netherlands: International Books.

Auken, S. [Danish Minister of Environment and Energy] (1996). *Notes for an Intervention at the Official Opening of the First Meeting of the Open-Ended Ad Hoc Working Group on Biosafety 22 July 1996*, Aarhus, Denmark.

Bacon, F. (1620). *The New Organon*, New York: F.H. Anderson (ed. 1960).

Barns, I. (1996). "Manufacturing Consensus?: Reflections on the UK National Consensus Conference on Plant Biotechnology", *Science as Culture* 5 (23), 199-216.

Barnthouse, L.W. (1992). "The Role of Models in Ecological Risk Assessment: A 1990's Perspective", *Environmental toxicology and chemistry 11* (12), 1751-1760.

Bazin, M.J. and Lynch, J.M. (eds.) (1994). *Environmental Gene Release: Models, Experiments and Risk Assessment*, London: Chapman and Hall (OECD).

Beatty, J. (1982). "The insights and oversights of molecular genetics: the place of the evolutionary perspective", *PSA 1980, volume 1*, East Lansing, Mich.: Philosophy of Science Association, 341-355.

Beck, U. (1992). *Risk Society, Towards a New Modernity*, London: SAGE.

Bell, J. (1999). "GM Foods Turn Political Hot Potato", *Seedling*, March, 8-11.

Berg, P. *et al.* (1975). "Asilomar Conference on Recombinant DNA Molecules", *Science* 188, 931-935.

Bergelson, J. (1994). "Changes in fecundity do not predict invasiveness: A model study of transgenic plants", *Ecology* 75 (1), 249-252.

Bergelson, J., Purrington, C.B. and Wichmann, G. (1998). "Promiscuity in transgenic plants", *Nature* 395 (September 3), 25.

Bergmans, J.E.N. (1992). "Horizontal Gene Transfer in Micro-Organisms: The Role of Plasmids and Transposons", in: J. Weverling and P. Schenkelaars (eds.), *Ecological Effects of Genetically Modified Organisms*, Arnhem: Netherlands Ecological Society, 81-91.

Bergmans, J.E.N. (1995). "What can we learn from experience with releases from GMOs? What approach should we take in the future?", in: CCRO (Coordination Commission Risk-Assessment Research), *Unanswered Safety Questions when Employing GMOs*, Overschild, the Netherlands: CCRO, 23-25.

Bergmans, J.E.N. and Middelhoven, W.J. (1998). "Nu toch gemeenschappelijk standpunt over toelaatbaarheid transgene A. tumefaciens", *BioNieuws* 8 (25 april), 2.

Beringer, J.E. and Bale, M.J. (1988). "The Release of Genetically Engineered Plants and Microorganisms", *Journal of Chemical Technology and Biotechnology* 43 (4), 273-278.

Bernhardt, M., Weber, B. and Tappeser, B. (1991/1994). *Gutachten zur biologischen Sicherheit bei der Nutzung der Gentechnik*, Freiburg: ÖKO-Institut.

Beusmann, V. (1995). "Diskurse um ökologische Implikationen der Freisetzung trangener Pflanzen als Teil von Technikfolgenabschätzungen (TA)", in: S. Albrecht and V. Beusmann (eds.), *Ökologie transgener Nutzpflanzen*, Frankfurt/New York: Campus Verlag, 21-40.

Beijersbergen, A. (1993). *Trans-Kingdom Promiscuity – Similarities between* T-DNA *Transfer by Agrobacterium Tumefaciens and Bacterial Conjugation*, Leiden.

Beijersbergen, A., Dulk-Ras, A.D., Schilperoort, R.A. and Hooykaas, P.J.J. (1992). "Conjugative Transfer by the Virulence System of Agrobacterium tumefaciens", *Science* 256, 1324-1327.

Bijman, W.J. and Lotz, L.A.P. (1996). *Transgene Herbicideresistente Rassen*, Den Haag: Ministerie van LNV.

Böger, P. (1994). *Mögliche pflanzenphysiologische Veränderungen in herbizidresistenten und transgenen Pflanzen und durch den Kontakt mit Komplementärherbiziden*, Berlin: Wissenschaftszentrum für Sozialforschung.

Boyce, N. (1997). "What's in a name...", *New Scientist* (26 July), 24.

Brandle, J.E., McHugh, S.G., James, L., Labbe, H. and Miki, B.L. (1995). "Instability of Transgene Expression in Field Grown Tobacco Carrying the csr-1 Gene for Sulphonylurea Herbicide Resistance", *Bio/Technology* 13 (September), 994-998.

Brandon, R.N. (1990). *Adaptation and Environment*, Princeton, New Jersey: Princeton University Press.

Brandon, R.N. (1992). "Environment", in: E.F. Keller and E.A. Lloyd (eds.), *Keywords in Evolutionary Biology*, Cambridge, MA: Harvard University Press, 81-86.

Braun, P.W. (1996). "Influence of Transgenes on Coevolutionary Processes", in: J. Tomiuk, A. Sentker, and K. Wöhrmann (eds.), *Transgenic Organisms – Biological and Social Implications*, Birkhäuser-Verlag, 99-111.

Breckling, B. (1993). *Naturkonzepten und Paradigmen in der Ökologie. Einige Entwicklungen*, Berlin: Wissenschaftszentrum für Sozialforschung.

Brill, W.J. (1985). "Safety concerns and genetic engineering in agriculture", *Science* 227 (25 January), 381-384.

Brock, Th.D. (1985). "Procaryotic Population Ecology", in: H.O. Halvorson, D. Pramer and M. Rogul (eds.), *Engineered Organisms in the Environment: Scientific Issues*, Washington, DC: American Society for Microbiology, 176-179.

Broer, I. (1995). "Folgenforschung an transgenen Pflanzen: Ein Beitrag zur Technikfolgenabschätzung", in: S. Albrecht and V. Beusmann (eds.), *Ökologie transgener Nutzpflanzen*, Frankfurt/New York: Campus Verlag, 99-109.

Broer, I. and Pühler, A. (1994). *Stabilität von* HR-*Genen in transgenen Pflanzen und ihrer spontaner horizontaler Gentransfer auf andere Organismen*, Berlin: Wissenschaftszentrum für Sozialforschung.

Brooks, H. (1984). "The Resolution of Technically Intensive Public Policy Disputes", *Science, Technology and Human Values* 9 (1), 39-50.

Brunt, J. van (1987). "Environmental Release: A Portrait of Opinion and Opposition", *Bio/Technology* 5, 559-663.

Buhl-Mortensen, L. (1996). "Type-II Statistical Errors in Environmental Science and the Precautionary Principle", *Marine Pollution Bulletin* 32 (7), 528-531.

Bunders, J. and Radder, H. (1995). "The Appropriate Realization of Agricultural Biotechnology", in: T.B. Mepham, G.A. Tucker and J. Wiseman (eds.), *Issues in Agricultural Bioethics*, Nottingham: Nottingham University Press, 177-204.

Cambrosio, A., Limoges, C. and Hoffman, E. (1992). "Expertise as a Network: A Case Study after the Controversies over Deliberate Release", in: N. Stehr and R. Ericson (eds.), *The Culture and Power of Knowledge*, De Gruiter, 341-361.

Canter Cremers, H.C.J. and Groot, H.F. (1991a). *Survival of E.coli K12 on laboratory coats made of 100% cotton*, Bilthoven: RIVM.

Canter Cremers, H.C.J. and Groot, H.F. (1991b). *The mailing of genetically modified microorganisms: A field survey*, Bilthoven: RIVM.

Carson, R. (1963). *Silent Spring*, Harmondsworth: Penguin.

Carter, J. (1998). "Who's Afraid of Genetic Engineering", *The New York Times*, August 26.

CEC [Commission of the European Communities] (1990). "Council Directive 90/220/ EEC on the deliberate release into the environment of genetically modified organisms", *Official Journal of the European Communities*, 15-27.

CEC [Commission of the European Communities] (1993). *Final Sectorial Meeting on Biosafety and First Sectorial Meeting on Microbial Ecology (Granada, October 24-27, 1993)*, Brussels: DG XII, Science, Research and Development.

CCRO [Coordination Commission Risk-Assessment Research] (ed.) (1995). *Unanswered Safety Questions when Employing GMOs*, Overschild, the Netherlands: CCRO.

Chargaff, E. (1987). "Engineering a Molecular Nightmare", *Nature* 327 (21 May), 199-200.

Chesson, J. (1990). "Data Requirements for Environmental Risk Assessment", in: J.J. Marois and G. Bruening (eds.), *Risk Assessment in Agricultural Biotechnology: Proceedings of the International Conference (Davis, California,* 1988), Oakland: Division of Agriculture and Natural Resources, University of California, 164-167.

Colborn, Th., Dumanoski, D. and Myers, J.P. (1996). *Our Stolen Future*, London: Abacus.

Collingridge, D. (1980). *Social Control of Technology*, Milton Keynes: The Open University Press.

Collingridge, D. and Earthy, M. (1990). "Science under Stress, Crisis in Neo-Darwinism", *History and Philosophy of the Life Sciences* 12, 3-26.

Collingridge, D. and Reeve, C. (1986). *Science Speaks to Power – The Role of Experts in Policy Making*, London: Frances Pinter.

Colwell, R.K. (1985). "Biological Responses to Perturbation: Genome to Ecosystem", in: H.O. Halvorson, D. Pramer and M. Rogul (eds.), *Engineered Organisms in the Environment: Scientific Issues*, Washington, DC: American Society for Microbiology, 230-232.

Colwell, R.K. (1988). "Ecology and biotechnology: Expectations and outliers", in: J. Fiksel and V.T. Covello (eds.), *Risk analysis approaches for environmental releases of genetically engineered organisms*, NATO Advanced Research Science Institute Series, Volume F, Berlin: Springer-Verlag.

Colwell, R.K. et al. (1987). "Response to the Office of Science and Technology Policy Notice 'Coordinated Framework for the Regulation of Biotechnology'", *Bulletin of the Ecological Society of America* 68, 16-23.

Condit, R. (1991). "Models for the Population Dynamics of Transposable Elements in Bacteria", in: L.R. Ginzburg (ed.), *Assessing Ecological Risks of Biotechnology*, Boston: Butterworth-Heinemann, 151-171.

Crawley, M.J. (1993). "Arm-Chair Risk Assessment (letter to editor)", *Bio/Technology 11.*

Crawley, M.J. (1994). "Long term ecological impacts of the release of genetically modified organisms", in: P.J. van der Meer and J.P.M. Schenkelaars (eds.), *PAN European Conference on the Long-term Ecological Impacts of Genetically Modified Organisms, Working Document 14 March*, Voorschoten, the Netherlands, 31-50.

Crawley, M.J., Hails, R.S., Rees, M., Kohn, D. and Buxton, J. (1993). "Ecology of Transgenic Oilseed Rape in Natural Habitats", *Nature* 363, 620-623.

Daele, W. van den (1992a). "Scientific Evidence and the Regulation of Technical Risks: Twenty Years of Demythologizing the Experts", in: N. Stehr and R. Ericson (eds.), *The Culture and Power of Knowledge*, De Gruiter, 323-340.

Daele, W. van den (1992b). *The Research Program of the Section "Norm-building and Environment"*, Berlin: Wissenschaftszentrum für Sozialforschung.

Daele, W. van den (1993). "Das falsche Signal zur falschen Zeit", *Politische Ökologie* 12 (35), 65.

Daele, W. van den (1994). *Technology Assessment as a Political Experiment – Discursive Procedure for the Technology Assessment of the Cultivation of Crop Plants with Genetically Engineered Herbicide Resistance*, Berlin: Wissenschaftszentrum für Sozialforschung.

Daele, W. van den, Pühler, A., Sukopp, H., Bora, A. and Döbert, R. (1994). *Bewertung und Regulierung von Kulturpflanzen mit gentechnisch erzeugter Herbizidresistenz (HR-Technik)*, Berlin: Wissenschaftszentrum für Sozialforschung.

Daele, W. van den, Pühler, A., Sukopp, H., Broer, I., Bora, A., Döbert, R., Neubert, S. and Siewert, V. (1994). *Argumentationen des TA-Verfahrens: Horizontaler Gentransfer aus Transgenen Pflanzen*, Berlin: Wissenschaftszentrum für Sozialforschung.

Daele, W. van den, Pühler, A. Sukopp, H. (1996). *Grüne Gentechnik im Widerstreit. Modell einer partizipativen Technikfolgenabschätzung zum Einsatz transgener herbizidresistenter Pflanzen*, Weinheim, BRD: VCH.

Daly, J.C. and Trowell, S. (1996). "Biochemical Approaches to the Study of Ecological Genetics: The Role of Selection and Gene Flow in the Evolution of Insecticide Resistance", in: W.O.C. Symondson and J.E. Liddell (eds.), *The Ecology of Agricultural Pests – Biochemical Approaches*, London: Chapman and Hall, 73-92.

Damme, J.M.M. van (1992). "Hybridisation between Wild and Transgenic Plants", in: J. Weverling and P. Schenkelaars (eds.), *Ecological Effects of Genetically Modified Organisms*, Arnhem: Netherlands Ecological Society, 81-91.

Davies, J. (1994). "Inactivation of Antibiotics and the Dissemination of Resistance Genes", *Science* 264, 375-382.

Davis, B.D. (1987). "Bacterial Domestication: Underlying Assumptions", *Science* 235 (13 March), 1329-1335.

Davis, B.D. (1989). "Evolutionary Principles and the Regulation of Engineered Bacteria", *Genome* 31, 864-869.

Day, M. and Fry, J.C. (1992). "Gene transfer in the environment: Conjugation", in: J.C. Fry and M.J. Day (eds.), *Release of Genetically Engineered and Other Micro-Organisms*, Cambridge, UK: Cambridge University Press, 40-53.

Dickson, D. (1995). "Biosafety Code Gathers Pace through Bilateral Agreements", *Nature* 377, 94.

Dommelen, A. van (1995). "Quality of Risk Assessment: Artificial and Fundamental Controversies", in: R. von Schomberg (ed.), *Contested Technology – Ethics, Risk and Public Debate*, Tilburg: Int. Centre for Human and Public Affairs, 193-207.

Dommelen, A. van (1996a). "The Impact of Background Models on the Quality of Risk Assessment as Exemplified by the Discussion on Genetically Modified Organisms", in: A. van Dommelen (ed.), *Coping with Deliberate Release – The Limits of Risk Assessment*, Tilburg/Buenos Aires: Int. Centre for Human and Public Affairs, 47-62.

Dommelen, A. van (1996b). "Milieucrisis en Methodologie: Over de Kwaliteit van Wetenschap", in: R. von Schomberg (red.), *Het Discursieve Tegengif – De Sociale en Ethische Aspecten van de Ecologische Crisis*, Kampen: Kok Agora, 157-186.

Dommelen, A. van (ed.) (1996c). *Coping with Deliberate Release – The Limits of Risk Assessment*, Tilburg/Buenos Aires: Int. Centre for Human and Public Affairs.

Dommelen, A. van (1998). "Useful Models for Biotechnology Hazard Identification: What is This Thing Called 'Familiarity'?", in: P. Wheale *et al.* (eds.), *The Social Management of Genetic Engineering*, Aldershot: Ashgate, 219-236.

Doucet, P. and Sloep, P. (1993). *Mathematical Models in the Life Sciences*, Chisester: Ellis Horwood.

Doyle, J. (1986). *Altered Harvest*, New York: Penguin.

Doyle, J.D., Stotzky, G., McClung, G. and Hendricks, C.W. (1995). "Effects of Genetically Engineered Microorganisms on Microbial Populations and Processes in Natural Habitats", *Advances in Applied Microbiology* 40, 237-287.

Doyle, J.J. and Persley, G.J. (1996). *Enabling the Safe Use of Biotechnology – Principles and Practice*, Washington: Worldbank.

Dunster, H.J. (1994). "Risk Assessment", in: M.J. Bazin and J.M. Lynch (eds.), *Environmental Gene Release: Models, Experiments and Risk Assessment*, London: Chapman and Hall (OECD), 139-148.

Egziabher, T.B.G. *et al.* (1995). *The Need for Greater Regulation and Control of Genetic Engineering – A Statement by Scientists Concerned about Current Trends in the New Biotechnology*, Penang, Malaysia: Third World Network.

Elsas, J.D. van (1995). "Preparation of protocols and decision trees for field trials with genetically engineered microorganisms. Key factors for decision trees", in: CCRO (ed.), *Unanswered Safety Questions when Employing GMOs*, Overschild, the Netherlands: CCRO, 123-127.

EPA [Environmental Protection Agency] (1994). "Microbial Products of Biotechnology; Proposed Regulation Under the Toxic Substances Control Act", *Federal Register* 59, 45600.

Fedoroff, N.V. (1991). "Maize Transposable Elements", *Perspectives in Biology and Medicine* 35 (Autumn), 2-19.

Fincham, J.R.S. and Ravetz, J.R. (1991). *Genetically engineered organisms, benefits and risks*, Buckingham: Open University Press.

Fitter, A., Perrins, J. and Williamson, M. (1990). "Weed probability challenged", *Bio/Technology* 8, 473.

Friedt, W. and Ordon, F. (1996). "Modern *versus* Classical Plant Breeding Methods – Efficient Synergism or Competitive Antagonism?", in: J. Tomiuk, A. Sentker and K. Wöhrmann (eds.), *Transgenic Organisms – Biological and Social Implications*, Birkhäuser-Verlag, 163-179.

Fry, J.C. and Day, M.J. (1990). "Plasmid Transfer and the Release of Genetically Engineered Bacteria in Nature: A Discussion and Summary", in: J.C. Fry and M.J. Day (eds.), *Bacterial Genetics in Natural Environments*, London: Chapman and Hall, 243-250.

Fry, J.C. and Day, M.J. (1992). *Release of Genetically Engineered and Other Micro-Organisms*. Cambridge, UK: Cambridge University Press.

Gabriel, W. (1993). "Technologically Modified Genes in Natural Populations: Some Skeptical Remarks on Risk Assessment from the View of Population Genetics", in: K. Wöhrmann and J. Tomiuk (eds.), *Transgenic Organisms: Risk-Assessments of Deliberate Release*, Basel: Birkhäuser, 109-116.

Gabriel, W. and Lynch, M. (1992). "The Selective Advantage of Reaction Norms for Environmental Tolerance", *Journal of Evolutionary Biology* 5, 41-59.

Gassen, H.G., Sachse, G., Stollwerk, J. and Zinke, H. (1991). *Gutachten zur Biologischen Sicherheit bei der Nutzung der Gentechnik – für das Büro für Technikfolgen-Abschätzung des Deutschen Bundestages* (*TAB*), Bonn, Darmstadt: Institut für Biochemie.

GAO [US General Accounting Office] (1988). *Biotechnology: Managing Risks of Field Testing Genetically Engineered Organisms*, Washington, DC: Government Printing Office.

Gibbs, W.W. (1997). "Plantibodies – Human antibodies produced by field crops enter clinical trials", *Scientific American* (November), 23.

Glandorf, D.C.M., Bakker, P.A.H.M. and van Loon, L.C. (1997). "Influence of the Production of Antibacterial and Antifungal Proteins by Transgenic Plants on the Saprophytic Soil Microflora", *Acta Botanica Neerlandica* 46 (1), 85-104.

Gliddon, C. (1994). "The Impact of Hybrids Between Genetically Modified Crop Plants and their Related Species: Biological Models and Theoretical Perspectives", *Molecular Ecology* 3 (1), 41-44.

Goodwin, B. (1996). "Species as Natural Kinds that Express Distinctive Natures: The Case for a Moratorium on Deliberate Release", in: A. van Dommelen (ed.), *Coping with Deliberate Release – The Limits of Risk Assessment*, Tilburg/Buenos Aires: Int. Centre for Human and Public Affairs, 73-78.

Gore, A. (1992). "Seeds of Privation", *Earth in the Balance, ecology and the human spirit*, Boston: Houghton Mifflin Company, 126-144.

Haefner, J.W. (1996). *Modeling Biological Systems – Principles and Applications*, New York: Chapman and Hall.

Hempel, C.G. (1966). *Philosophy of Natural Science*, Englewood Cliffs: Prentice-Hall.

Hengeveld, R. (1992). "Cause and Effect in Natural Invasions", in: J. Weverling and P. Schenkelaars (eds.), *Ecological Effects of Genetically Modified Organisms*, Arnhem: Netherlands Ecological Society, 29-43.

Hengeveld, R. (1994). "Assessing Invasion Risk", in: P.J. van der Meer and J.P.M. Schenkelaars (eds.), *PAN European Conference on the Long-term Ecological Impacts of Genetically Modified Organisms, Working Document 14 March*, Voorschoten, the Netherlands, 67-82.

Hey, E. (1996). "Interpreting the Precautionary Principle", *International environmental affairs* 8 (1), 95-98.

Hill, A. and Michael, M. (1998). "Engineering Acceptance: Representation of "the public" in debates on biotechnology", in: P. Wheale *et al.* (eds.), *The Social Management of Genetic Engineering*, Aldershot: Ashgate, 201-217.

Hill, J. (1994). "The Precautionary Principle and Release of Genetically Modified Organisms (GMOs) to the Environment", in: T. O'Riordan and J. Cameron (eds.), *Interpreting the Precautionary Principle*, London: Earthscan, 172-182.

Ho, M.W. (1997). "The Unholy Alliance", *The Ecologist* 27 (July/August).

Ho, M.W. and Tappeser, B. (1997). "Potential Contributions of Horizontal Gene Transfer to the Transboundary Movement of Living Modified Organisms Resulting from Modern Biotechnology", in: K.J. Mulongoy (ed.), *Transboundary Movement of Living Modified Organisms Resulting from Modern Biotechnology: Issues and Opportunities for Policy-Makers*, Conches-Geneva, Switzerland: Int. Academy of the Environment, 171-193.

Holmes, M.T. and Ingham, E.R. (1994). "The Effects of Genetically Engineered Microorganisms on Soil Foodwebs", *Supplement to Bulletin of Ecological Society of America 75/2*, Abstracts of the 79th Annual ESA Meeting: Science and Public Policy, Knoxville, Tennessee, 7-11 August 1994.

Holmes, M.T., Ingham, E.R., Doyle, J.D. and Hendricks, C.W. (1999). "Effects of Klebsiella planticola SDF20 on soil biota and wheat growth in sandy soil", Applied Soil Ecology 11 (1), 67-78.

HRH the Prince of Wales (1998). "Seeds of Disaster", *The Ecologist* 28 (5), 252-253.

Hueting, R. and Reijnders, L. (1998). "Sustainability is an objective concept", *Ecological Economics*, 27 (October), 139-147.

Independent Group of Scientific and Legal Experts on Biosafety (1996). *Biosafety – Scientific Findings and Elements of a Protocol*, Penang, Malaysia: Third World Network.

Ingham, E., Holmes, M., Johnston, R. and Tuininga, A. (1995). *Biosafety Regulation: A Critique of Existing Documents*, Edmonds, Washington: The Edmonds Institute.

Istock, C.A. (1991). "Genetic Exchange and Genetic Stability in Bacterial Populations", in: L.R. Ginzburg (ed.), *Assessing Ecological Risks of Biotechnology*, Boston: Butterworth-Heinemann, 123-149.

Jäger, M. and Tappeser, B. (1996). "Politics and Science in Risk Assessment", in: A. van Dommelen (ed.), *Coping with Deliberate Release – The Limits of Risk Assessment*, Tilburg/Buenos Aires: Int. Centre for Human and Public Affairs, 63-72.

Jasanoff, S. (1987). "Contested Boundaries in Policy-Relevant Science", *Social Studies of Science* 17, 195-230.

Jasanoff, S. *et al.* (eds.) (1995). *Handbook of Science and Technology Studies*, Thousand Oaks, CA: SAGE.

Jukes, T.H. (1988). "Hazards of Biotechnology: Facts and Fancy", *Journal of Chemical Technology and Biotechnology* 43 (4), 245-255.

Käppeli, O. and Auberson, L. (1997). "The Science and Intricacy of Environmental Safety Evaluations", *Trends in Biotechnology* 15, 342-349.

Kareiva, P. (1993). "Transgenic Plants on Trial", *Nature* 363, 580-581.

Kareiva, P., Manasse, R. and Morris, W. (1991). "Using Models to Integrate Data from Field Trials and Estimate Risks of Gene Escape and Gene Spread", *Biological Monitoring of Genetically Engineered Plants and Microbes*, ARS, 31-42.

Kareiva, P. and Parker, I. (1994). *Environmental Risks of Genetically Engineered Organisms and Key Regulatory Issues*, Greenpeace International.

Kareiva, P., Parker, I.M. and Pascual, M. (1996). "Can we use experiments and models in predicting the invasiveness of genetically engineered organisms?", *Ecology* 77 (6), 1670-1674.

Kasanmoentalib, S. (1996). "Deliberate Release of Genetically Modified Organisms: Applying the Precautionary Principle", in: A. van Dommelen (ed.), *Coping with Deliberate Release – The Limits of Risk Assessment*, Tilburg/Buenos Aires: Int. Centre for Human and Public Affairs, 137-146.

Kay, L.E. (1993). *The Molecular Vision of Life – Caltech, the Rockefeller Foundation, and the rise of the new biology*, New York: Oxford University Press.

Keeler, K.H. (1985). "Implications of Weed Genetics and Ecology for the Deliberate Release of Genetically Engineered Crop Plants", *Recombinant DNA Technical Bulletin* 8 (4), 165-172.

Keeler, K.H. (1989). "Can Genetically Engineered Crops Become Weeds", *Bio/Technology* 7 (11), 1134-1139.

Keller, E.F. (1990). "Physics and the Emergence of Molecular Biology: A History of Cognitive and Political Synergy", *Journal of the History of Biology* 23, 389-409.

Kiernan, V. (1996). "Yes, we have vaccinating bananas", *New Scientist* (21 September), 6.

Köhler, W. and Braun, P. (1995). "Populationsgenetische und statistische Aspekte der Begleitforschung", in: S. Albrecht and V. Beusmann (eds.), *Ökologie transgener Nutzpflanzen*, Frankfurt/New York: Campus Verlag, 163-181.

Kokjohn, T.A. and Miller, R.V. (1992). "Gene transfer in the environment: Transduction", in: Fry, J.C. and Day, M.J. (eds.), *Release of Genetically Engineered and Other Micro-Organisms*, Cambridge, UK: Cambridge University Press, 54-81.

Kollek, R. (1992). *Kommentar zum "Gutachten zur biologischen Sicherheit ... ' des Instituts für Biochemie/TH Darmstadt (Dezember 1991) und zum "Gutachten zur "biologischen" Sicherheit ... " des -Instituts Freiburg (November 1991)*, Bonn: Büro für Technikfolgen-Abschätzung der DB.

Kollek, R. (1993a). "The Gene Concept: Historical Disputes and Their Relation to Current Controversies in Genetic Engineering", Paper given at the *XIX International Congress of History of Science, 22-29 July, Zaragoza, Spain.*

Kollek, R. (1993b). "Controversies about Risks and their Relation to Different Paradigms in Biological Research", in: R. von Schomberg (ed.), *Science, Politics and Morality – Scientific Uncertainty and Decision Making*, Dordrecht: Kluwer Academic Publishers, 27-42.

Kollek, R. (1995). "The Limits of Experimental Knowledge: A Feminist Perspective on the Ecological Risk of Genetic Engineering", in: V. Shiva and I. Moser (eds.), *Biopolitics: A Feminist and Ecological Reader on Biotechnology*, London: Zed Books, 95-111.

Kooijman, S.A.L.M. (1993). *Dynamic Energy Budgets in Biological Systems – Theory and Applications in Ecotoxicology*, Cambridge, UK: Cambridge University Press.

Krimsky, S. (1991). *Biotechnics and Society: The Rise of Industrial Genetics*, New York: Praeger.

Krimsky, S. (1996). "Risk Assessment of Genetically Engineered Microorganism: From Genetic Reductionism to Ecological Modeling", in: A. van Dommelen (ed.), *Coping with Deliberate Release – The Limits of Risk Assessment*, Tilburg/Buenos Aires: Int. Centre for Human and Public Affairs, 33-45.

Krimsky, S. and Wrubel, R. (1993). *Agricultural Biotechnology: An Environmental Outlook*, Medford, MA: Tufts University.

Krimsky, S. and Wrubel, R. (1996). *Agricultural Biotechnology and the Environment*, Urbana and Chicago: University of Illinois Press.

Krimsky, S., Wrubel, R.P., Naess, I.G., Levy, S.B., Wetzler, R.E. and Marshall, B. (1995). "Standardized Microcosms in Microbial Risk Assessment", *Bioscience* 45 (9), 590-599.

Kuhn, T.S. (1963). *The Structure of Scientific Revolutions*, Chicago: University of Chicago Press.

Kuijen, C.J. van (1992). "Environmental Risks Policy on Genetically Modified Organisms in the Netherlands and Europe", in: J. Weverling and P. Schenkelaars (eds.), *Ecological Effects of Genetically Modified Organisms*, Arnhem: Netherlands Ecological Society, 23-27.

Lake, G.J. (1989). "The STOA Experiment in the European Parliament", *Energy Policy* (June), 284-288.

Lake, G.J. (1991). "Scientific uncertainty and political regulation: European legislation on the contained use and deliberate release of genetically modified (micro) organisms", *Project Appraisal* 6 (1), 7-15.

Lau, C. (1992). "Social Conflicts about the Definition of Risks: The Role of Science", in: N. Stehr and R. Ericson (eds.), *The Culture and Power of Knowledge*, De Gruiter, 235-248.

Leatherdale, W.H. (1974). *The Role of Analogy, Model and Metaphor in Science*, Amsterdam: North-Holland.

Lenski, R.E. (1993). "Evaluating the Fate of Genetically Modified Organisms in the Environment: Are they Inherently less fit?", *Experientia* 49, 201-209.

Lenteren, J.C. van (1992). "Insect Invasions: Origins and Effects", in: J. Weverling and P. Schenkelaars (eds.), *Ecological Effects of Genetically Modified Organisms*, Arnhem: Netherlands Ecological Society, 59-80.

Levidow, L. (1995). "Whose Ethics for Agricultural Biotechnology", in: V. Shiva and I. Moser (eds.), *Biopolitics: A Feminist and Ecological Reader on Biotechnology*, London: Zed Books, 175-190.

Levidow, L. *et al.* (1996). "Bounding the Risk Assessment of a Herbicide-Tolerant Crop", in: A. van Dommelen (ed.), *Coping with Deliberate Release – The Limits of Risk Assessment*, Tilburg/Buenos Aires: Int. Centre for Human and Public Affairs, 81-102.

Levin, S.A. (1991). "An Ecological Perspective", in: B. Davis (ed.), *The Genetic Revolution, Scientific Prospects and Public Perceptions*, Baltimore and London: The Johns Hopkins University Press, 45-59.

Levins, R. and Lewontin, R. (1985). *The Dialectical Biologist*, Cambridge, MA: Harvard University Press.

Lewontin, R.C. (1992). "The Dream of the Human Genome", *The New York Review of Books* XXXIX (May 28), 31-40.

Lindow, S.E. (1985). "Ecology of *Pseudomonas syringae* Relevant to the Field Use of Ice-minus Deletion Mutants Constructed In Vitro for Plant Frost Control", in: H.O. Halvorson, D. Pramer and M. Rogul (eds.), *Engineered Organisms in the Environment: Scientific Issues*, Washington, DC: American Society for Microbiology, 23-35.

Lindow, S.E. (1990). "Environmental Use of Genetically Engineered Organisms", in: E. Heseltine (ed.), *Advances in Biotechnology*, Stockholm: Swedish Council for Forestry and Agricultural Research/Swedish Recombinant DNA Advisory Committee, 101-114.

Longino, H.E. (1990). *Science as Social Knowledge: Values and Objectivity in Scientific Inquiry*, Princeton, NJ: Princeton University Press.

Lorenz, M.G. and Wackernagel, W. (1996). "Mechanisms and Consequences of Horizontal Gene Transfer in Natural Bacterial Populations", in: J. Tomiuk, A. Sentker and K. Wöhrmann (eds.), *Transgenic Organisms – Biological and Social Implications*, Birkhäuser-Verlag, 45-57.

Luhmann, N. (1986). *Ökologische Kommunikation, kann die moderne Gesellschaft sich auf ökologische Gefährdungen einstellen?*, Opladen: Westdeutscher Verlag.

MacKenzie, D. (1996). "Doctors farm fish for insulin", *New Scientist* (16 November), 20.

MacKenzie, D. (1997). "Modified maize faces widening opposition", *New Scientist* (15 February), 10.

Maessen, G.D.F. (1997). "Genomic Stability and Stability of Expression in Genetically Modified Plants", *Acta Botanica Neerlandica* 46 (1), 3-24.

Manasse, R.S. (1992). "Ecological risks of transgenic plants: effects of spatial dispersion on gene flow", *Ecological Applications* 2, 431-438.

Mantegazzini, M.C. (1986). *The Environmental Risks from Biotechnology*, London: Frances Pinter.

McIntosh, R.P. (1985). *The Background of Ecology – concept and theory*, New York: Cambridge University Press.

McIntosh, R.P. (1987). "Pluralism in Ecology", *Annual Review of Ecology and Systematics* 18, 321-341.

McNally, R. and Wheale, P. (1998). "The Consequences of Modern Genetic Engineering: Patents, "Nomads" and the "Bio-Industrial Complex"", in: P. Wheale *et al.* (eds.), *The Social Management of Genetic Engineering*, Aldershot: Ashgate, 303-330.

Meer, P.J. van der (1994). "Introduction to the conference", in: P.J. van der Meer and J.P.M. Schenkelaars (eds.), *PAN European Conference on the Long-term Ecological Impacts of Genetically Modified Organisms, Working Document 14 March, Voorschoten, the Netherlands, 11-14.*

Melcher, R. and Barrett, A. (1999). "Fields of Genes: Reengineered crops will change the way the world feeds, clothes, and heals itself", *Business Week*, 12 April: 46-52.

Mellon, M. and Rissler, J. (1995). "Transgenic Crops: USDA Data on Small-Scale Tests Contribute Little to Commercial Risk Assessment", *Bio/Technology* 13 (January), 96.

Mergeay, M., De Rore, H., Top, E., Springael, D., Höfte, M., van der Lelie, D., Dijkmans, R. and Verstraete, W. (1994). "Plasmid Biology and Risk Assessment: Elements for Modelling Approach of Plasmid-mediated Gene Release in Soil Environments", in: M.J. Bazin and J.M. Lynch (eds.), *Environmental Gene Release: Models, Experiments and Risk Assessment*, London: Chapman and Hall (OECD), 67-75.

Metz, P.L.J. and Nap, J.P. (1997). "A Transgene-centred Approach to the Biosafety of Transgenic Plants: Overview of Selection and Reporter Genes", *Acta Botanica Neerlandica* 46 (1), 25-50.

Michaels, P.J. and Knappenberger, P.C. (1996). "The United Nations Intergovernmental Panel on Climate Change and the Scientific "Consensus" on Global Warming", in: J. Emsley (ed.), *The Global Warming Debate*, London: European Science and Environment Forum, 158-178.

Middelhoven, W.J. (1997). "Standpunt van COGEM in flagrante tegenspraak met eigen richtlijnen", *BioNieuws* 7 (22 februari), 4.

Middelhoven, W.J. (1997). "Bestuur COGEM gaat niet op de hoofdzaken in", *BioNieuws* 7 (5 april), 2.

Mieth, D. (1993). "The Release of Microorganisms – Ethical Criteria", in: K. Wöhrmann and J. Tomiuk (eds.), *Transgenic Organisms: Risk-Assessments of Deliberate Release*, Basel: Birkhäuser, 245-256.

Miller, H.I. (1995). "Unscientific Regulation of Agricultural Biotechnology: Time to Hold the Policymakers Accountable?", *Trends in Biotechnology* 13, 123-125.

Miller, H.I. (1996). "Biosafety Regulations", *Nature* 379, 13.

Miller, H.I. (1997). *Policy Controversy in Biotechnology: An Insider's View*, Austin, TX: Landes Company.

Miller, H.I. and Gunary, D. (1993). "Serious Flaws in the Horizontal Approach to Biotechnology Risk", *Science* 262 (3 December), 1500-1501.

Miller, H.I., Huttner, S.L. and Altman, D.W. (1996). "Misunderstanding Risk", *Bio/Technology* 14 (January), 9.

Miller, R.M. (1993). "Nontarget and Ecological Effects of Transgenically Altered Disease Resistance in Crops – Possible Effects on the Mycorrhizal Symbiosis", *Molecular Ecology* 2, 327-335.

Miller, R.V. (1998). "Bacterial Gene Swapping in Nature", *Scientific American* (January), 47-51.

Moser, I. (1995). "Mobilizing Critical Communities and Discourses on Modern Biotechnology", in: V. Shiva and I. Moser (eds.), *Biopolitics: A Feminist and Ecological Reader on Biotechnology*, London: Zed Books, 1-24.

Nap, J.P., Bijvoet, J., Stiekema, W.J. (1992). "Biosafety of Kanamycin-resistant Plants", *Transgenic Research* (1), 239-249.

NAS [National Academy of Sciences] (1987). Committee on the Introduction of Genetically Engineered Organisms into the Environment, *Introduction of Recombinant DNA-engineered Organisms in the Environment: Key Issues*, Washington, DC: National Academy Press.

NAS [National Academy of Sciences] (1989). *Field Testing Genetically Modified Organisms: Framework for Decisions*, Washington, DC: National Academy Press.

Nelkin, D. (ed.) (1992). *Controversy. Politics of Technical Decisions*, London: Sage Focus (3rd edition).

Nester, E.W. Yanofsky, M.F. and Gordon, M.P. (1985). "Molecular Analysis of Host Range of Agrobacterium tumefaciens", in: H.O. Halvorson, D. Pramer and M. Rogul (eds.), *Engineered Organisms in the Environment: Scientific Issues*, Washington, DC: American Society for Microbiology, 191-196.

Nestle, M. (1996). "Allergies to Transgenic Foods – Questions of Policy", *The New England Journal of Medicine* 334, 726-728.

NSNE [Netherlands Society for Nature and Environment] (1997). "Review of Developments in Scientific Views on Potentially Adverse Effects of Genetically Modified Organisms: Implications for Regulatory Decisions", in: Netherlands Society for Nature and Environment (NSNE) (eds.), *Inventory of views of non-governmental organizations on the risk evaluation of genetically modified organisms in six cases*, 35-54.

OECD [Organization for Economic Cooperation and Development] (1993a). *Safety Considerations for Biotechnology: Scale-up of Crop Plants*, Paris: OECD.

OECD [Organization for Economic Cooperation and Development] (1993b). *Traditional crop breeding practices: An historical overview to serve as a baseline for assessing the role of modern biotechnology*, Paris: OECD.

OECD [Organization for Economic Cooperation and Development] (1995). *Safety Considerations for Biotechnology: Scale-up of Micro-organisms as Biofertilizers*, Paris: OECD.

OECD [Organization for Economic Cooperation and Development] (1998). *21st Century Technologies*, Paris: OECD.

O'Riordan, T. and Jordan, A. (1995). "The Precautionary Principle in Contemporary Environmental Politics", *Environmental Values* 4, 191-212.

PEER [Public Employees for Environmental Responsibility] (1995). *Genetic Genie – The Premature Commercial Release of Genetically Engineered Bacteria*, Washington, DC: PEER.

Perrow, Ch. (1984). *Normal Accidents: Living with high-risk technologies*, New York: Basic Books.

Peterman, R.M. (1990). "Statistical Power Analysis can Improve Fisheries Research and Management", *Canadian Journal for Fisheries and Aquatic Sciences*, 2-15.

Peterman, R.M. and M'Gonigle, M. (1992). "Statistical Power Analysis and the Precautionary Principle", *Marine pollution bulletin* 24, 231-234.

Pols, B. (1998). "Biotechnologie in Amerika en Europa: Wachten op de eerste echte blockbuster", *NRC Handelsblad* (5 juni).

Popper, K.R. (1968/1934). *The Logic of Scientific Discovery*, New York: Harper and Row.

Postman, N. (1989). "Dwangbuis", in: M. Schwarz and R. Jansma (eds.), *De Technologische Cultuur*, Amsterdam: De Balie, 27-31.

Potthast, T. (1996). "Trangenic Organisms and Evolution: Ethical Impliations", in: J. Tomiuk, A. Sentker and K. Wöhrmann (eds.), *Transgenic Organisms – Biological and Social Implications*, Basel: Birkhäuser-Verlag, 227-240.

Prins, H., van Wieren, S. and Olff, H. (1998). "Wetenschappelijke integriteit en natuurbescherming gaan wel degelijk samen", *BioNieuws* 8 (28 maart), 2.

Raybould, A.F. and Gray, A.J. (1994). "Will Hybrids of Genetically Modified Crops Invade Natural Communities?", *Trends in Ecology and Evolution* 9 (3), 85-89.

Regal, P.J. (1985). "The Ecology of Evolution: Implications of the Individualistic Paradigm", in: H.O. Halvorson, D. Pramer and M. Rogul (eds.), *Engineered Organisms in the Environment: Scientific Issues*, Washington, DC: American Society for Microbiology, 11-19.

Regal, P.J. (1986). "Models of Genetically Engineered Organisms and Their Ecological Impact", in: H.A. Mooney and J.A. Drake (eds.), *Ecology of Biological Invasions of North America and Hawaii*, New York, NY: Springer-Verlag, 111-129.

Regal, P.J. (1987). "Safe and Effective Biotechnology: Mobilizing Scientific Expertise", in: J.R. Fowle III (ed.), *Application of Biotechnology: Environmental and Policy Issues*, Boulder, CO: Westview Press, 145-164.

Regal, P.J. (1988). "The Adaptive Potential of Genetically Engineered Organisms in Nature", in: J. Hodgson and A.M. Sugden (eds.), *Planned release of genetically engineered organisms. Trends in biotechnology / Trends in ecology and evolution special publication*, Cambridge, UK: Elsevier 6 (4), s36-s38.

Regal, P.J. (1990). "Gene Flow and Adaptability in Transgenic Agricultural Organisms: Long Term Risks and Overview", in: J.J. Marois and G. Bruening (eds.), *Risk Assessment in Agricultural Biotechnology: Proceedings of the International Conference (Davis, California, 1988)*, Oakland: Division of Agriculture and Natural Resources, University of California, 102-110.

Regal, P.J. (1990). *The Anatomy of Judgment*, Minneapolis: University of Minnesota Press.

Regal, P.J. (1993). "The True Meaning of "Exotic Species" as a Model for Genetically Engineered Organisms", *Experientia* 49 (3), 225-234.

Regal, P.J. (1994). "Scientific Principles for Ecologically Based Risk Assessment of Transgenic Organisms", *Molecular Ecology* 3, 5-13.

Regal, P.J. (1995a). "International Biosafety: A Global Imperative", *The Scientist* 9 (21), 13.

Regal, P.J. (1995b). *Biosafety Protocol: The Need for an International Binding Regulatory Mechanism*, Edmonds, Washington: The Edmonds Institute.

Regal, P.J. (1996). "Metaphysics in Genetic Engineering: Cryptic Philosophy and Ideology in the "Science" of Risk Assessment", in: A. van Dommelen (ed.), *Coping with Deliberate Release – The Limits of Risk Assessment*, Tilburg/Buenos Aires: Int. Centre for Human and Public Affairs, 15-32.

Regal, P.J. (1997). "The Geography of Risk: Special Concerns for Insular Ecosystems and for Centres of Crop Origins and Genetic Diversity", in: K.J. Mulongoy (ed.), *Transboundary Movement of Living Modified Organisms Resulting from Modern Biotechnology: Issues and Opportunities for Policy-Makers*, Conches-Geneva, Switzerland: Int. Academy of the Environment, 159-169.

Rehbinder, E. (1994). *Rechtsprobleme gentechnisch veränderter herbizidresistenter Pflanzen*, Berlin: Wissenschaftszentrum für Sozialforschung.

Rehmann-Sutter, C. (1993). "Nature in the Laboratory – Nature as Laboratory. Considerations about the Ethics of Release Experiments", *Experientia* 49, 190-200.

Rehmann-Sutter, C. and Vatter, A. (1996). "Risk Communication and the Ethos of Democracy", in: A. van Dommelen (ed.), *Coping with Deliberate Release – The Limits of Risk Assessment*, Tilburg/Buenos Aires: Int. Centre for Human and Public Affairs, 207-226.

Rehmann-Sutter, C. (1998). "Involving others: Towards an ethical concept of risk", *Risk* 9 (2), 117-136.

Reijnders, L. (1992). "Negative Effects and Invasions", in: J. Weverling and P. Schenkelaars (eds.), *Ecological Effects of Genetically Modified Organisms*, Arnhem: Netherlands Ecological Society, 101-103.

Rip, A. (1992). "Expert Advice and Pragmatic Rationality", in: N. Stehr and R. Ericson (eds.), *The Culture and Power of Knowledge*, De Gruiter, 363-379.

Ruef, K. (1997). *The Private Eye : Looking/Thinking by Analogy – A Guide to Developing the Interdisciplinary Mind.* The Private Eye Project.

Ruesink, J.L., Parker, I.M., Groom, M.J. and Kareiva, P.M. (1995). "Reducing the Risks of Nonindigenous Species Introductions – Guilty Until Proven Innocent", *BioScience* 45 (7), 465-477.

Sayre, P.G. and Miller, R.V. (1991). "Bacterial mobile genetic elements: Importance in assessing the environmental fate of genetically engineered sequences", *Plasmid* 26, 151-171.

Schell, J.S. (1994). "Plant Biotechnology: State of the art in developed countries and relevant safety considerations", in: P.J. van der Meer and J.P.M. Schenkelaars (eds.), *PAN European Conference on the Long-term Ecological Impacts of Genetically Modified Organisms, Working Document 14 March, Voorschoten, the Netherlands, 17-24.*

Schellekens, H. and Bergmans, J. (1997). "Maatregelen moeten in proportie zijn met gevaar van verspreiding", *BioNieuws* 7 (22 februari), 4.

Schmidt, K. (1995). "Whatever Happened to the Gene Revolution", *New Scientist* (7 January), 21-25.

Schmitt, J.J. (1995). "Institutionalisierung von Sicherheitsforschung? Wissenschaftliche und politische Aspekte zur Organisation in Deutschland", in: S. Albrecht and V. Beusmann (eds.), *Ökologie transgener Nutzpflanzen*, Frankfurt/New York: Campus Verlag, 215-228.

Schomberg, R. von (1993). "Controversies and Political Decision Making", in: R. von Schomberg (ed.), *Science, Politics and Morality – Scientific Uncertainty and Decision Making*, Dordrecht: Kluwer Academic Publishers, 7-26.

Schomberg, R. von (1995). *Der rationale Umgang mit Unsicherheit – Die Bewältigung von Dissens und Gefahren in Wissenschaft, Wissenschaftspolitik und Gesellschaft*, Frankfurt am Main: Europäischer Verlag der Wissenschaften.

Schomberg, R. von (1997). *Argumentatie in de Context van een Wetenschappelijke Controverse – Een analyse van de discussie over de introductie van genetisch gemodificeerde organismen in het milieu*, Delft: Eburon.

Schomberg, R. von (1998a). *An appraisal of the working in practice of directive 90/220/EEC on the deliberate release of genetically modified organisms*, Brussels: STOA.

Schomberg, R. von (1998b). "Democratising the Policy Process for the Environmental Release of Genetically Engineered Organisms", in: P. Wheale *et al.* (eds.), *The Social Management of Genetic Engineering*, Aldershot: Ashgate, 237-248.

Schütte, G. (1995). "Pflanzenschutz mit Hilfe der Gentechnik am Beispiel einer virus-resistenten Zuckerrübensorte – Denkbare Folgen, Risiken und offene Fragen", in: S. Albrecht and V. Beusmann (eds.), *Ökologie transgener Nutzpflanzen*, Frankfurt/ New York: Campus Verlag, 183-213.

Schweder, T. (1996). "The Two Strategies of Biological Containment of Genetically Engineered Bacteria", in: J. Tomiuk, A. Sentker and K. Wöhrmann (eds.), *Transgenic Organisms – Biological and Social Implications*, Birkhäuser-Verlag, 113-125.

Segerstråle, U. (1990). "Negotiating "Sound Science": Expert Disagreement about Release of Genetically Engineered Organisms", *Politics and the Life Sciences* 8 (2), 221-231.

Segerstråle, U. (1992). "Reductionism, 'Bad Science' and Politics", *Politics and the Life Sciences*. 11, 2 (August), 199-214.

Segerstråle, U. (1993). "Bringing the Scientist Back In: The Need for an Alternative Sociology of Scientific Knowledge", in: Th. Brante, S. Fuller and W. Lynch (eds.), *Controversial Science*, New York: SUNY, 57-82.

Sentker, A. (1996). "Genetic Engineering and the Press – Public Opinion *versus* Published Opinion", in: J. Tomiuk, A. Sentker and K. Wöhrmann (eds.), *Transgenic Organisms – Biological and Social Implications*, Birkhäuser-Verlag, 241-254.

Sharples, F.E. (1983). "Spread of Organisms with Novel Genotypes: Thoughts from an Ecological Perspective", *Recombinant DNA Technology Bulletin* 6 (June), 43-56.

Sharples, F.E. (1987). "Application of Introduced Species Models to Biotechnology Assessment", in: J.R. Fowle III (ed.), *Application of Biotechnology: Environmental and Policy Issues*, Boulder, CO: Westview Press, 93-98.

Sharples, F.E. (1991). "Ecological Aspects of Hazard Identification for Environmental Uses of Genetically Engineered Organisms", in: M.A. Levin and H.S. Strauss (eds.), *Risk Assessment in Genetic Engineering, Environmental Release of Organisms*, New York: McGraw-Hill, 18-31.

Shiva, V. and Moser, I. (eds.) (1995). *Biopolitics: A Feminist and Ecological Reader on Biotechnology*, London: Zed Books.

Shiva, V. (1995). "Epilogue: Beyond Reductionism", in: V. Shiva and I. Moser (eds.), *Biopolitics: A Feminist and Ecological Reader on Biotechnology*, London: Zed Books, 267-284.

Sloep, P.B. (1993). "Methodology Revitalized?", *British Journal for the Philosophy of Science* 44, 231-249.

Sloep, P.B. (1994). "The Impact of "Sustainability" on the Field of Environmental Science", in: G. Skirbekk (ed.), *The Notion of Sustainability and its Normative Implications*, Oslo: Scandinavian University Press, 29-55.

Smalla, K. and Van Elsas, J.D. (1996). "Monitoring Genetically Modified Organisms and their Recombinant DNA in Soil Environments", in: J. Tomiuk, A. Sentker and K. Wöhrmann (eds.), *Transgenic Organisms – Biological and Social Implications*, Birkhäuser-Verlag, 127-146.

Snow, C.P. (1961). *Science and Government*, Cambridge, MA: Harvard University Press.

Steen, W.J. van der (1993). *A Practical Philosophy for the Life Sciences*, New York: SUNY.

Steen, W.J. van der, and Kamminga, H. (1991). "Laws and natural history in biology", *British Journal for the Philosophy of Science* 42, 445-467.

Stephenson, J.R. and Warnes, A. (1996). "Release of Genetically Modified Microorganisms into the Environment", *Journal of Chemical Technology and Biotechnology* 65 (1), 5-14.

Sterling, W.L. (1990). "Appropriate Monitoring Strategies for Non-Target Populations", in: J.J. Marois and G. Bruening (eds.), *Risk Assessment in Agricultural Biotechnology: Proceedings of the International Conference (Davis, California, 1988)*, Oakland: Division of Agriculture and Natural Resources, University of California, 153-163.

Stewart, G.J. (1992). "Gene transfer in the environment: Transformation", in: J.C. Fry and M.J. Day (eds.), *Release of Genetically Engineered and Other Micro-Organisms*, Cambridge, UK: Cambridge University Press, 82-93.

Stotzky, G. (1990). "Ecological Considerations Related to the Release of Genetically Engineered Microorganisms to the Environment", in: E. Heseltine (ed.), *Advances in Biotechnology*, Stockholm: Swedish Council for Forestry and Agricultural Research/ Swedish Recombinant DNA Advisory Committee, 145-157.

Stotzky, G., Broder, M.W., Doyle, J.D. and Jones, R.A. (1993). "Selected Methods for the Detection and Assessment of Ecological Effects Resulting from the Release of Genetically Engineered Microorganisms to the Terrestrial Environment", *Advances in Applied Microbiology* 38, 1-98.

Strauss, H.S. (1991). "Lessons from Chemical Risk Assessment", in: M.A. Levin and H.S. Strauss (eds.), *Risk Assessment in Genetic Engineering, Environmental Release of Organisms*, New York: McGraw-Hill, 297-318.

Sukopp, U. and Sukopp, H. (1993). "Das Modell der Einführung und Einbürgerung nicht einheimischer Arten – Ein Beitrag zur Diskussion über die Freisetzung gentechnisch veränderter Kulturpflanzen", *GAIA* 2 (5), 267-288.

Sukopp, U. and Sukopp, H. (1994). *Ökologische Lang-Zeiteffekte der Verwilderung von Kulturpflanzen*, Berlin: Wissenschaftszentrum für Sozialforschung.

Sukopp, H. and Sukopp, U. (1995). "Ökologische Modelle in der Begleitforschung zur Freisetzung transgener Kulturpflanzen", in: S. Albrecht and V. Beusmann (eds.), *Ökologie transgener Nutzpflanzen*, Frankfurt/New York: Campus Verlag, 41-64.

Szanto, T.R. (1993). "Value Communities in Science: The Recombinant DNA Case", in: Th. Brante, S. Fuller and W. Lynch (eds.), *Controversial Science*, New York: SUNY, 241-263.

Szybalski, W. (1985a). "Genetic engineering in agriculture (letter)", *Science* 229 (12 July), 112-115.

Szybalski, W. (1985b). "Early Warning Principle Offsets the Need for Regulation of the Recombinant DNA Technique", *BioEssays* 2, 147.

Tait, J. and Levidow, L. (1992). "Proactive and Reactive Approaches to Regulation: The Case of Biotechnology", *Futures* (April), 219-231.

Tangley, L. (1985). "Releasing engineered organisms in the environment – Assessing the risks of genetically engineered organisms will require closer collaboration between molecular biologists and ecologists", *BioScience* 35 (8), 470-473.

Tappeser, B. (1994). *Kommentargutachten zur studie: "Stabilität von HR-Genen in transgenen Pflanzen und ihrer spontaner horizontaler Gentransfer auf andere Organismen"*, Berlin: Wissenschaftszentrum für Sozialforschung.

Tenenbaum, D. (1996). "Weeds From Hell", *Technology Review* (August/September), 33-40.

Thompson, P.B. (1987). "Agricultural Biotechnology and the Rhetoric of Risk: Some Conceptual Issues", *Environmental Professional* 9, 316-326.

Tiedje, J.M., Colwell, R.K., Grossman, Y.L., Hodson, R.E., Lenski, R.E., Mack, R.N. and Regal, P.J. (1989). "The planned introduction of genetically engineered organisms: ecological considerations and recommendations", *Ecology* 70 (2), 298-315.

Toulmin, S. (1990). *Cosmopolis – The Hidden Agenda of Modernity*, Chicago: The University of Chicago Press.

Trepl, L. (1987). *Geschichte der Ökologie, vom 17. Jahrhundert bis zur Gegenwart*, Frankfurt am Main: Athenäum.

Tversky, A. and Kahneman, D. (1974). "Judgment under Uncertainty: Heuristics and Biases", *Science* 185, 1124-1131.

UCS [Union of Concerned Scientists] (1995). "Why UCS Resists Bt Crops", *The Gene Exchange* 5-6 (4-1), 2/15.

UNCED [United Nations Conference on Environment and Development] (1992). "Environmentally Sound Management of Biotechnology", *Agenda 21: Programme of Action for Sustainable Development*, Rio de Janeiro, Brazil: UNCED, 136-146.

UNEP [United Nations Environmental Programme] (1994). *Convention on Biological Diversity – Text and Annexes*, Switzerland: UNEP.

UNEP [United Nations Environmental Programme] (1996). *International Technical Guidelines for Safety in Biotechnology.*

Verhoog, H. (1993). "Biotechnology and Ethics", in: Th. Brante, S. Fuller and W. Lynch (eds.), *Controversial Science*, New York: SUNY, 83-106.

Watson, J.D. and Tooze, J. (1981). *The DNA Story, a documentary history of gene cloning*, San Francisco: W.H. Freeman.

Weber, B. (1994). *Evolutionsbiologische Argumente in der Risikodiskussion am Beispiel der transgenen herbizidresistenten Pflanzen*, Berlin: Wissenschaftszentrum für Sozialforschung.

Weber, B. (1995). "Überlegungen zur Aussagekraft von Risikoforschung zur Freisetzung transgener Pflanzen", in: S. Albrecht and V. Beusmann (eds.), *Ökologie transgener Nutzpflanzen*, Frankfurt/New York: Campus Verlag, 111-126.

Weber, B. (1996). ""Normalisierung" durch Vergleich? – Zur Bedeutung der Risikodebatte nach Abschluß des WZB-Verfahrens zur gentechnisch erzeugten Herbizidresistenz", *TA-Datenbank-Nachrichten* 5 (4), 18-25.

Weele, C. van der (1999). *Images of Development: Environmental Causes in Ontogeny*, New York: SUNY.

Weidhase, R.A. (1994). *Kommentar zum Gutachten: 'Mögliche pflanzenphysiologische Veränderungen in herbizidresistenten und transgenen Pflanzen und durch den Kontakt mit Komplementärherbiziden'*, Berlin: Wissenschaftszentrum für Sozialforschung.

Weinberg, A.M. (1972). "Science and Trans-Science", *Minerva* 10, 209-222.

Weinberg, A.M. (1985). "Science and Its Limits: The Regulator's Dilemma", *Issues in Science and Technology* 2, 209-222.

Weizsäcker, C. von (1995). "Error-friendliness and the Evolutionary Impact of Deliberate Release of GMOs", in: V. Shiva and I. Moser (eds.), *Biopolitics: A Feminist and Ecological Reader on Biotechnology*, London: Zed Books, 112-120.

Weizsäcker, C. von (1996). "Lacking Scientific Knowledge or Lacking the Wisdom and Culture of Not-Knowing?", in: A. van Dommelen (ed.), *Coping with Deliberate Release – The Limits of Risk Assessment*, Tilburg/Buenos Aires: Int. Centre for Human and Public Affairs, 195-206.

Wheale, P. and McNally, R. (1988). *Genetic Engineering: Catastrophe or Utopia?*, Hemel Hempstead: Harvester Wheatsheaf.

Wheale, P. and McNally, R. (1994). "What 'Bugs' Genetic Engineers About Bioethics – The Consequences of Genetic Engineering as Post-Modern Technology", in: A. Dyson and J. Harris (eds.), *Ethics and biotechnology*, London: Routlegde, 179-201.

Williamson, M. (1992). "Environmental risks from the release of genetically modified organisms (GMOs) – The need for molecular ecology", *Molecular Ecology* 1, 3-8.

Williamson, M. (1993). "Invaders, weeds, and the risks from GMOs", *Experientia* 49, 219-224.

Williamson, M. (1994). "Community Response to Transgenic Plant Release: Predictions from British Experience of Invasive Plants and Feral Crop Plants", *Molecular Ecology* 3 (1), 75-79.

Williamson, M., Perrins, J. and Fitter, A. (1990). "Releasing Genetically Engineered Plants: Present Proposals and Possible Hazards", *Trends in Ecology and Evolution* 5 (12 December), 417-419.

Wimsatt, W.C. (1984). "Reductionistic Research Strategies and Their Biases in the Units of Selection Controversy", in: R.N. Brandon and R.M. Burian (eds.), *Genes, Organisms, Populations: Controversies*, Cambridge MA: MIT Press, 90-108.

Wittgenstein, L. (1965). *The Blue and Brown Books*, New York: Harper and Row.

Wittgenstein, L. (1971). *Philosophische Untersuchungen*, Frankfurt am Main: Suhrkamp.

Wright, K. (1986). "New Regulations in Dispute", *Nature* 323 (2 October), 387.

Wright, S. (1986). "Molecular Biology or Molecular Politics? The Production of Scientific Consensus on the Hazards of Recombinant DNA Technology", *Social Studies of Science* 16, 593-620.

Wright, S. (1994). *Molecular Politics – Developing American and British Regulatory Policy for Genetic Engineering, 1972-1982*, Chicago.

Wrubel, R.P., Krimsky, S. and Wetzler, R.E. (1992). "Field Testing Transgenic Plants, an Analysis of the US Department of Agriculture's Environmental Assessments", *BioScience* 42 (4), 280-289.

WWF [World Wildlife Fund] (1995). *Genetic Engineering – Examples of Ecological Effects and Inherent Uncertainties.*

Young, F.E. and Miller, H.I. (1988). "'Old' Biotechnology to 'New' Biotechnology: Continuum or Disjunction?", in: J. Fiksel and V.T. Covello (eds.), *Safety Assurance for Environmental Introductions of Genetically-Engineered Organisms (NATO Advanced Research Workshop, Rome, Italy, 1987)*, New York: Springer-Verlag, 13-33.

Zadoks, J.C. and De Wit, P.J.G.M. (1992). "Invasiveness of Plant Pathogenic Micro-organisms", in: J. Weverling and P. Schenkelaars (eds.), *Ecological Effects of Genetically Modified Organisms*, Arnhem: Netherlands Ecological Society, 13-21.

Zannoni, L. (1995). "The quest for harmonization of regulatory oversight for biotechnology, the OECD experience...", in: CCRO (ed.), *Unanswered Safety Questions when Employing GMOs*, Overschild, the Netherlands: CCRO, 35-39.

SUMMARY

Hazard Identification of Agricultural Biotechnology: Finding Relevant Questions

— Chapter 1 —

In the introductory Chapter 1, the context and purpose of this research is explicated. The general problem that is studied in this book is the occurrence of controversy in science for policy. The specific focus of this study is on *controversies over biosafety assessment of genetic engineering.* The present research is an attempt to develop a fundamental and pragmatic approach to this pressing problem. Since controversies in applied science will typically involve more than one scientific discipline and will often affect scientists as well as nonscientists, it is a requirement that the framework of analysis is both accessible for diverse parties in the debate and practically applicable to the issues of discussion.

New technological possibilities pose new challenges to our foresight of their possible consequences. The potential impact of genetic engineering is a concern that merits attention. Scientist are not in agreement about the consequences that genetic engineering may have in terms of possible disturbances to ecosystems, the inadvertent escape of transgenic inserts to non-target organisms, and other undesirable effects to our health and our natural environment. Where scientists disagree, policy-makers find themselves in an uncertain position to take decisions about the future. The methodological status of scientific controversies in the context of policy-making needs to be clarified to allow rational and democratic public decisions.

The case study of the biosafety assessment of genetic engineering that is central throughout this book, has the status of an elaborate test case here. Other issues of environmental safety and the timely recognition of inadvertent effects can also be reconstructed in terms of impact assessment and the differences of opinion that exist about it within the realm of scientific expertise.

An important prerequisite for any debate is special concern for the proper definitions of the subjects of discussion. Without adequate scientific insights possible effects may go unnoticed for too long, as the history of environmental impact assessment has shown. In this respect, scientific models can be seen as the 'sensory organs of science' without which a possible problem will not be noticed in time. In the applied context of environmental impact assessment, scientific experts can be seen as 'hazard detectives'.

When scientists are in controversy about the adequate assessment of a possible environmental impact, this may be a threat to the objective of avoiding inadvertent effects. In such a situation, scientific expertise may turn into an obstacle for practical wisdom rather that an asset. The controversies in applied science may be hardened under the stress of social relevance and thereby may easily degenerate in 'endless technical debate'. One possible approach is to accept this mechanism of conflict and leave it to politicians and policy-makers to choose those scientific advisors that appear to them as presenting the most plausible claims. However, this would be a deflation of the meaning of scientific expertise and the status of scientific theory in society.

A possible loss of quality in the practical application of scientific expertise can be countered by an appropriate methodological analysis of scientific controversy in an applied context. Problem definitions that are at the basis of controversy in applied science may effectively be reconstructed in terms of the *research questions that are relevant to address the problem at hand.* If a particular problem of environmental impact assessment is not sufficiently specified in terms of the relevant questions to be addressed for a satisfactory understanding of the problem, then the result may in effect be an 'artificial controversy' rather than a 'fundamental controversy'. By comparing and evaluating relevant questions rather than conflicting claims, the focus in the controversy is shifted from the opposing arguments in the debate to the complementing 'windows of concern'. The latter focus is more in accordance with the purpose of environmental impact assessment.

The applicability of biosafety analogies such as 'traditional breeding' or 'exotic species', for example, can only be assessed in precisely defined terms, *i.e.*, in terms of the specific relevant quesions that need to be addressed in a given context. Both analogies represent a 'window of concern' that can offer useful insights in the nature of the problem of the biosafety assessment of genetic engineering. On the other hand, neither can make claim to privileged applicability. From this preliminary example, a generally applicable analytical framework is developed in Chapter 2.

— Chapter 2 —

To meet the scientific requirements for making an adequate biosafety assessment, we need an appropriate 'window of concern' for understanding the research problem. The required biosafety tests for assessing the possible effects of genetic engineering can be reconstructed as lists of research questions that need to be addressed to be able to support claims about expected impacts. The adequacy and reliability of any specific test and thereby the plausibility of any claim that will be based on the outcome of that test, will be decided by the specific research questions that together are constitutive of that test.

An important distinction for the purpose of biosafety assessment of genetic engineering is the one between *risk analysis* and *hazard identification*. Before we can make supported scientific claims about the chance of a possible impact to occur, we must identify the range of possible effects that could be expected. Genetically engineered *Klebsiella planticola*, for example, was not recognized as possibly hazardous in one testing experiment, whereas it was recognized as a threat to the environment in an alternative biosafety test in which additional questions were considered as possibly relevant for the purpose of hazard identification.

Scientific methodology can be reconstructed as the *art of asking the right questions* in view of any specific research problem and as a sufficient basis for any specific claim. For this purpose, the concept of a *set of relevant questions* (SRQ) is introduced and defined. As an example of how the methodology of science can be constitutive of a practical biosafety assessment, the misguided use of the concept of 'competitiveness' by officials of the US *Department of Agriculture* in relation to the environmental introduction of genetically engineered organisms (GEOs) is analysed.

Complexity reduction in the sense of reducing a research problem to its essence for a specific purpose of investigation, is a prerequisite for scientific research and must be accounted for in all cases. The presented methodological analysis in terms of the relevance of research questions for the specified purpose of research, clarifies the involved scientific *burden of proof* for making claims in the context of hazard identification, given the purpose of preventing harm. It implies that the design of an experiment will be decisive for the interpretation of its results and that 'empirical evidence' can only be evaluated against the background of the research questions that were considered relevant for the purpose of the investigation.

As an example, the use of the concept of 'pathogenicity' is analysed as a possible methodological pitfall giving rise to misguided plausibility and *artificial* controversy. Analysis of controversies in terms of relevant research questions that define the problem of study can constitute a breakthrough in seemingly irresolvable issues. It moves the debate from being right to being concerned, making the controversy productive again. The challenge to the scientific community becomes one of generating research questions and arguing about their relevance for the purpose of detecting possible inadvertent effects. This allows the involved experts to join forces for the common purpose of hazard idenfication, instead of being involved in unproductive artificial controversy about a concern with important implications for society.

— Chapter 3 —

In Chapter 3, the analytical framework as developed in Chapter 2 is put to the test in a practical situation and compared to an elaborate evaluative framework that was applied in Germany in a technology assessment (TA) with respect to transgenic herbicide resistance in plants. The challenge of this chapter is to see how the suggested approach could make a difference compared to existing approaches in a practical situation of controversy in applied science.

In the elaborate TA study, an important role was explicitly given to the scientific basis of the input in the debate. However, a methodological analysis of the procedure and its results shows that insufficient attention is given to an adequate use of the scientific concepts and to reaching clarity over the research problems that are the concern of the argumentation. The TA study is characterized as a "political experiment" by its coordinators, which illustrates the important social and political role of scientific expertise. In the end, however, the experiment led to the departure of a substantial part of the enrolled critical experts.

The project coordinators have stressed the supposed empirical nature of the conclusions that were reached, but at the same time they have failed to clarify the methodological and experimental basis of these empirical conclusions. Since this is an important prerequisite for avoiding artificial controversy, the methodological status of the "empirical conclusions" remains hidden in the debate for lack of a satisfactory analytical framework.

As an example, we can consider the unclarity about a proper interpretation of the concept of a 'phenotype', which is a fundamental methodological prerequisite for making empirical claims about a safety concern such as 'selective advantage' of herbicide resistance. Since the analysed TA procedure has not produced an adequate methodological framework, there is no basis for discussions about the appropriate burden of proof for comparing new technologies with earlier practices. To define 'similarity' of compared technologies or practices, the relevant aspects of study in a specific context must first be made explicit in terms of relevant research questions for a specified purpose of investigation.

The importance of considering the different contexts of research is illustrated by the 'clashes' between the different scientific disciplines of ecology and molecular biology. On the basis of a methodological analysis of the scientific controversies, we may conclude that the relevance of *both* questions at the biological level of organization that is studied by ecology *and* at the level of biological organization that is studied by molecular biology, must be argued for in view of the purpose of hazard identification. The practical application of the SRQ approach could have led the German TA study to different final conclusions and might have prevented the departure of critical scientific experts from the debate. The claim to being a participatory technology assessment has not been met, for want of a fundamental and pragmatic methodological analysis of the scientific controversies that were at stake.

— Chapter 4 —

In Chapter 4, the analytical framework as developed in Chapter 2 and applied in Chapter 3 is used to recognize and evaluate four categories of ongoing and recurring controversial issues throughout the general debate about the biosafety assessment of genetic engineering. It is demonstrated that they can be characterized as being artificially defined and therefore pose an unproductive burden to the debate. The productive pragmatics of the SRQ approach can help to bring back the debate to a constructive appreciation of the arguments that are put forward in view of the purpose of hazard identification. The presented analyses serve to demonstrate that the analytical approach developed in this study works for arguments and claims in the biosafety debate more generally and that individual claims or arguments can be analysed along its lines.

The contrasting opinions about the desirability of regulating either the *processes* of genetic engineering, or the *products* of genetic engineering are reconstructed in terms of the relevant research questions that are addressed by either of the policy options. Application of the SRQ approach makes it clear that *both* in relation to the products of genetic engineering *and* in relation to the processes of genetic engineering, the relevance of research questions must be argued for and comes with a burden of proof. This takes away the unproductive exclusiveness of the conflicting perspectives and opens up the constructive complementarity of the two general positions for the purpose of hazard identification. This is demonstrated by an analysis of the methodological unclarity about the notion of 'precision' and about the role of transposons in the context of this debate.

Next, the contrasting opinions about the desirability of regulating genetic engineering by *case-by-case review* or by *generic review* are analysed by application of the SRQ approach. In this context, the concept of 'adaptation' can be analysed as a possible methodological pitfall that may make the controversy artificial rather than fundamental. An additional difficulty that is analysed is the biological importance of the receiving ecosystem. Since it only makes sense to make this concern specific in terms of the relevant research questions, failure to do so is a recipe for artificial controversy. By taking the perspective of finding the relevant questions for hazard identification, it becomes clear that *both* the considerations implied in a case-by-case review *and* the considerations implied in a generic review may improve the general understanding for the purpose of hazard identification.

A special concern in debates about the biosafety assessment of genetic engineering is related to the hazard identification of genetically engineered *micro*organisms (GEMS). It is demonstrated by application of the SRQ approach that a methodological analysis of the ongoing controversies shows that at least part of the discussion is about artificial concerns. The plausibility of claims to the effect that genetically engineered microorganism have already developed in nature throughout the history of biological evolution or that GEMS will not be harmful and that 'no vacancies' exist in a specific ecosystem, can only be assessed by specifying the relevant research questions that need to be addressed in view of the purpose of hazard identification.

Generally important to the context of biosafety assessment is the concept of 'probability' in an evolutionary context. Application of the SRQ approach shows that the implied notion of 'independent chances' in biology must be specified by addressing the relevance of research questions in view of the purpose of study. As an example of an 'artificial controversy' about the probability of inadvertent effects in the context of the biosafety assessment of genetic engineering, the claim that there would be a mechanism of 'early warning' available or present in the biological world is analysed and reconstructed in terms of the questions that would be relevant to support it.

— Chapter 5 —

In the context of biosafety policy, the notion of 'familiarity' has been introduced in an attempt to pragmatically overcome the difficulty of finding a satisfactory scientific basis for biosafety assessment. This attempt is frustrated by the fact that an operational scientific definition of 'familiarity' is missing in this context.

In Chapter 5, it is demonstrated that the SRQ approach can be applied as a practical methodological basis for a scientific definition of the notion of 'familiarity'. Familiarity can only be made specific and useful in a context of scientific expertise for policy purposes by making it explicit in terms of the set of research questions that are considered relevant to claim sufficient acquaintance with the subject of investigation. Unspecified use of the notion invites the risk of creating an artificial understanding of what are the scientific requirements for making a reliable hazard identification and biosafety assessment. The suggested pragmatic approach allows for the design and organization of an international research project with the objective of specifying the scientific requirements for claiming familiarity in terms of the research questions that are considered relevant for a satisfactory investigation of the subject.

The notion of 'horizontal gene transfer' and its use in debates about the biosafety assessment of genetic engineering is analysed as an example to illustrate that 'familiarity' can only be a useful concept in the context of biosafety assessment if it is operationalized in terms of specific research questions. The SRQ approach allows for a *dynamic* and *modular* procedure in the developing definition of familiarity. Controversies that exist all over the world about the biosafety of genetic engineering can thus be made productive by interpreting them as sources for research questions that are possibly relevant for the given purpose. By focusing on the questions, differences of opinion may become fruitful rather than endless and frustrating.

The practical advantage of this approach is that it makes it possible to embark on the development, as an international process of joint forces, of a *dynamic* and *modular* catalog of research questions that may together define familiarity. Such a catalog may become a resource for the specification of notification and monitoring requirements such as specied by regulatory frameworks as *Annex II* of EU Council Directive 90/220. Thus, it may also help national advisory boards, such as the Dutch COGEM (*Commission for Genetic Modification*), for which at present it is not clear in some cases what would be a sufficient scientific ground to decide against the release of a permit for field testing or market application.

As a practical format for the development of such a catalog, we need a provisional definition of what constitutes a 'hazard' and thus what aspects of familiarity are required to be studied for the purpose of hazard identification. The definition that is suggested is: 'A 'GEO poses a *hazard*' if 'it carries an *Agent P*, which can impose an *Effect Q* which is considered undesirable in *Context R*, to an *Affected S*, via a *Mechanism T*, in an *Environment X*, as a consequence of *Application Z*'. This opens up seven predicates of possible relevance in the design of appropriate research programmes and monitoring strategies.

— Chapter 6 —

In Chapter 6, the consequences of the findings in this study for the practical status of scientific expertise are brought together and evaluated. The SRQ approach implies a fundamental responsibility for scientific experts.

If politicians with a democratic basis of legitimation decide not to address some possibly relevant question(s), then this is a political decision in its proper context. If scientists take such a decision, then this is a political decision in an improper context. The demarcation of scientific and *transscientific* questions in the context of biosafety assessment is a responsibility of scientists, since they are in the best position to judge which questions can (or could) and which questions cannot be answered by science.

Political or ethical evaluations cannot be made without a clear conception and grasp of the methodological status of the scientific claims that are involved in the evaluation. A scientific expert, *qua scientist*, has a responsibility to safeguard the quality and relevance of empirical data. Precautionary science dedicates special awareness to the limitations of its claims and attempts to recognize an existing lack of information with the help of critical scientific methodology. Without this methodological scrutiny, which constitutes the very essence of scientific research, there can hardly be a proper way to apply the internationally accepted *precautionary principle* and to interpret its normative evaluations.

Methodological pitfalls such as in the interpretation of statistical results that are produced for the purpose of hazard identification can only be recognized and avoided by the scientific experts who are involved with the challenge of biosafety assessment. An important task of science is to protect us from thinking that we know more than we actually do. In the epilogue of the study, the main conclusions of the research are brought together and are put in a perspective of the associated *processes of learning* that are a *conditio sine qua non* to meet the scientific requirements for a precautionary biosafety assessment of genetic engineering.

Gevaarherkenning van landbouw-gentechnologie: het vinden van relevante vragen

— Hoofdstuk 1 —

In het inleidende Hoofdstuk 1 worden de context en het doel van dit onderzoek uiteengezet. Het algemene onderwerp van dit boek is de status van *wetenschappelijke controverses aan de basis van beleidsbeslissingen*. De specifieke aandacht is in deze studie gericht op *controverses over de bioveiligheidsbeoordeling van gentechnologie.*

Dit onderzoek beoogt een fundamenteel *en* pragmatisch perspectief op deze nijpende problematiek te bieden. Omdat controverses in toegepaste wetenschap in de meeste gevallen raken aan meerdere wetenschappelijke disciplines en omdat in veel gevallen zowel wetenschappers als niet-wetenschappers betrokken zullen zijn bij het debat, is het van belang dat het gepresenteerde analyse-raamwerk tegelijk toegankelijk is voor diverse deelnemers in het debat en praktisch toepasbaar is op het onderwerp van discussie.

Nieuwe technologische mogelijkheden stellen nieuwe uitdagingen aan ons vermogen om de mogelijke gevolgen te voorzien. De potentiële effecten van gentechnologie vragen om zorgvuldige aandacht. Onder wetenschappers bestaat geen overeenstemming over de gevolgen die gentechnologie kan hebben ten aanzien van mogelijke verstoringen van ecosystemen, de onverhoopte overdracht van transgenen naar andere organismen en overige mogelijke effecten op onze gezondheid en onze natuurlijke omgeving.

Waar wetenschappers van mening verschillen, verkeren beleidsmakers en politici in een onzekere positie ten aanzien van beslissingen over de toekomst. De methodologische status van wetenschappelijke controverses in beleidscontexten vraagt om opheldering ter bevordering van een rationele en democratische besluitvorming.

De voorbeeldstudie van de bioveiligheidsbeoordeling van gentechnologie die centraal staat in dit boek, wordt hier gepresenteerd als een uitgebreide *test case* voor het ontwikkelde analyse-raamwerk. Andere kwesties die betrekking hebben op de beoordeling van milieueffecten of de tijdige signalering van ongewenste gevolgen van technologische toepassingen, kunnen eveneens worden geanalyseerd in termen van het belang van wetenschappelijke deskundigheid voor de mogelijkheid van effectbeoordelingen.

Een belangrijke voorwaarde voor ieder debat is voldoende aandacht voor het gebruik van heldere definities met betrekking tot het onderwerp van discussie. Zonder de benodigde wetenschappelijke inzichten kunnen mogelijke milieueffecten geruime tijd onopgemerkt blijven, zoals de geschiedenis van het milieubewustzijn laat zien. In deze context kunnen wetenschappelijk modellen worden beschouwd als de 'zintuigen van de wetenschap', zonder welke een mogelijk probleem niet op tijd gesignaleerd kan worden. In de toegepaste context van de beoordeling van milieueffecten kan de rol van wetenschappelijke deskundigen worden gezien als die van 'gevaar detectives'.

Wanneer wetenschappers van mening verschillen over een adequate beoordeling van mogelijke milieueffecten, dan kan dit een bedreiging vormen voor het oogmerk om ongewenste gevolgen te vermijden. In een dergelijke situatie kan de wetenschappelijke deskundigheid een obstakel gaan vormen voor praktische wijsheid in plaats van een belangrijk hulpmiddel.

Bestaande controverses in toegepaste wetenschap kunnen nog verscherpt worden onder druk van de maatschappelijke relevantie van het debat en kunnen aldus ontaarden in 'eindeloze technische discussies'. Een mogelijk antwoord op dit gevaar van wetenschappelijke conflictversterking is om het aan beleidsmakers en politici over te laten om díe wetenschappelijke adviseurs te kiezen die naar hun mening de meest plausibele standpunten innemen. Echter, dit zou neerkomen op een miskenning van de betekenis van wetenschappelijke deskundigheid en van de status van wetenschappelijke kennis in de samenleving.

Een mogelijk verlies aan kwaliteit in de praktische toepassing van wetenschappelijke deskundigheid kan worden bestreden door een adequate methodologische analyse van wetenschappelijke controverses in een toegepaste context. Probleemdefinities die ten grondslag liggen aan controverses in toegepaste wetenschap kunnen met effect worden gereconstrueerd in termen van de *onderzoeksvragen die relevant zijn voor een doelgerichte behandeling van het betreffende probleem.*

Wanneer een specifiek probleem ten aanzien van de beoordeling van mogelijke milieueffecten niet voldoende precies wordt omschreven in termen van de onderzoeksvragen die relevant zijn voor een adequaat begrip van dat probleem, dan kan dit resulteren in een 'kunstmatige controverse' onder de wetenschappelijke deskundigen in plaats van een 'fundamentele controverse'.

Door het vergelijken en evalueren van *relevante vraagstellingen* in plaats van *conflicterende beweringen*, kan het onderwerp van de controverse worden verplaatst van de tegengestelde argumentaties naar elkaar aanvullende 'vensters van zorg' ('windows of concern'). Een dergelijke aanvullende benadering is meer in overeenstemming met de doelstelling van een bioveiligheidsbeoordeling.

De toepasbaarheid van bioveiligheidsanalogieën zoals 'traditionele veredeling' en 'exotische introducties' kan bijvoorbeeld alleen worden beoordeeld in voldoende precies gedefinieerde termen, dwz. in termen van de specifieke relevante onderzoeksvragen die binnen een bepaalde context moeten worden gesteld. Beide analogieën vertegenwoordigen een 'venster van zorg' dat bruikbare inzichten kan bieden in de aard van het probleem van de bioveiligheidsbeoordeling van gentechnologie binnen de landbouwpraktijk. Tegelijkertijd kan geen van beide analogieën aanspraak maken op een omvattende toepasbaarheid voor dit doel. In het verlengde van dit inleidende voorbeeld wordt in Hoofdstuk 2 een algemeen toepasbaar analyse-raamwerk geprestenteerd.

— Hoofdstuk 2 —

Voor een wetenschappelijke beoordeling van bioveiligheid is het noodzakelijk om een 'venster van zorg' te ontwikkelen dat voldoende licht werpt op het onderzoeksprobleem. De benodigde bioveiligheidstesten voor de beoordeling van mogelijke effecten van gentechnologie kunnen worden gereconstrueerd als reeksen van onderzoeksvragen die gesteld moeten worden om beweringen over mogelijk te verwachten effecten te kunnen onderbouwen.

De bruikbaarheid en betrouwbaarheid van een test, en daarmede ook van de beweringen die worden gebaseerd op de resultaten van die test, worden bepaald door de specifieke onderzoeksvragen die tezamen de test vormen.

Een belangrijk onderscheid voor het doel van de bioveiligheidsbeoordeling van gentechnologie is dat tussen *risicoanalyse* ('risk analysis') en *gevaarherkenning* ('hazard identification'). Voordat we onderbouwde wetenschappelijke beweringen kunnen doen over de kans dat een mogelijk effect zich voordoet, moeten we eerst vaststellen welke mogelijke effecten kunnen worden verwacht. Genetisch veranderde *Klebsiella planticola* werd bijvoorbeeld niet herkend als mogelijk gevaarlijk in het ene testexperiment, terwijl het wel als zodanig tevoorschijn kwam uit een alternatieve bioveiligheidstest waarin aanvullende onderzoeksvragen waren opgenomen als mogelijk relevant voor het doel van gevaarherkenning.

Wetenschappelijke methodologie kan worden gereconstrueerd als de *kunst van het vragen stellen* met betrekking tot een specifiek onderzoeksprobleem als voldoende basis voor specifieke beweringen daaromtrent. Ten behoeve van deze invalshoek wordt het begrip *verzameling van relevante vragen* (set of relevant questions, SRQ) ingevoerd en gedefinieerd. Als een voorbeeld van hoe de algemene methodologie van wetenschappelijk onderzoek ten grondslag ligt aan praktische bioveiligheidsbeoordeling, wordt getoond hoe een misvatting omtrent het begrip 'competitie' bij medewerkers van het Amerikaanse Ministerie van Landbouw kon leiden tot een inadequate beoordeling van de mogelijke effecten van het gebruik van genetisch veranderde organismen in het milieu.

Complexiteitsreductie, in de betekenis van het terugvoeren van een onderzoeksprobleem tot de essentie vanuit het oogpunt van het onderzoeksdoel, is noodzakelijk voor wetenschappelijk onderzoek en moet in alle gevallen verantwoord worden. De gepresenteerde methodologische analyse in termen van de relevantie van onderzoeksvragen voor een specifiek onderzoeksdoel, brengt helderheid ten aanzien van de wetenschappelijke *bewijslast* voor beweringen ten aanzien van gevaarherkenning. De analyse impliceert dat het ontwerp van een experiment beslissend zal zijn voor de interpretatie van de onderzoeksresultaten en dat 'empirisch bewijs' alleen beoordeeld kan worden tegen de achtergrond van de onderzoeksvragen die als relevant voor het onderzoeksdoel werden beschouwd.

Als voorbeeld wordt het begrip 'pathogeniciteit' geanalyseerd als een mogelijke methodologische valkuil die aanleiding kan geven tot ongefundeerde plausibiliteitsclaims en tot kunstmatige controverses. Door controverses te analyseren in termen van de relevante onderzoeksvragen die tezamen het onderzoeksprobleem definiëren, kan een doorbraak worden gerealiseerd in schijnbaar onverzoenbare tegenstellingen. De uitdaging voor de wetenschappelijke gemeenschap binnen de toegepaste context van bioveiligheidsbeoordeling van gentechnologie wordt erdoor gericht op de relevantie van onderzoeksvragen ten behoeve van het doel om mogelijke ongewenste effecten op tijd te herkennen. Dit stelt de deskundigen in staat om de krachten te bundelen voor het gemeenschappelijke doel van gevaarherkenning en om onproductieve kunstmatige controverses te vermijden.

— Hoofdstuk 3 —

In hoofdstuk 3 wordt het analyse-raamwerk dat werd ontwikkeld in hoofdstuk 2 op de proef gesteld in een praktische situatie en vergeleken met de benadering die werd toegepast in een uitgebreid technologisch aspectenonderzoek (TA) dat in Duitsland werd uitgevoerd met betrekking tot transgene herbicideresistentie van planten.

De uitdaging van dit hoofdstuk is om te demonstreren hoe de hier voorgestelde benadering een verschil kan maken ten opzichte van bestaande benaderingen van controverses in toegepaste wetenschap.

In de onderzochte TA-studie werd expliciet een belangrijke rol toegekend aan de wetenschappelijke basis van de standpunten in het debat. Een methodologische analyse van de procedure en de resultaten van de TA-studie laat echter zien dat onvoldoende aandacht werd gegeven aan een adequaat gebruik van het betreffende wetenschappelijke begrippenkader en aan een adequate definitie van de ter discussie staande onderzoeksproblemen.

De Duitse TA-studie is door de procesbegeleiders gekarakteriseerd als een "politiek experiment", waardoor het sociale en politieke belang van de inbreng van wetenschappelijke deskundigheid wordt onderstreept. Het experiment leidde echter tot een voortijdig vertrek van een aanzienlijk deel van de deelnemende kritische deskundigen. De procesbegeleiders in deze TA-studie hebben het veronderstelde empirische karakter van de bereikte conclusies benadrukt, maar zijn er tegelijkertijd niet in geslaagd om opheldering te geven over de methodologische en experimentele basis van die empirische conclusies. Omdat dit een belangrijke voorwaarde is voor het vermijden van kunstmatige controverses, blijft de methodologische status van de "empirische conclusies" verborgen in het debat bij gebrek aan een adequaat analyse-raamwerk.

Als voorbeeld wordt de in de TA-studie bestaande onduidelijkheid over een juiste interpretatie van het begrip 'fenotype' geanalyseerd. Duidelijkheid hieromtrent is een fundamentele methodologische voorwaarde voor het onderbouwen van empirische beweringen over een bioveiligheidsaspect als het 'selectieve voordeel' van herbicide-resistentie. Omdat de geanalyseerde TA-procedure een adequaat methodologisch raamwerk ontbeert, is er ook geen basis voor een bevredigende discussie over de verdeling van de wetenschappelijke bewijslast ten aanzien van vergelijkingen tussen het gebruik van nieuwe technologieën enerzijds en ervaringen met reeds bestaande praktijken anderzijds. Om de 'vergelijkbaarheid' van verschillende technologieën of praktijken te definiëren, moeten de van belang zijnde aspecten vooraf expliciet worden gemaakt in termen van de onderzoeksvragen die als relevant worden beschouwd voor het specifieke onderzoeksdoel.

Het belang van aandacht voor verschillende onderzoekscontexten wordt geïllustreerd door de 'botsingen' tussen de verschillende wetenschappelijke disciplines van ecologie en moleculaire biologie. Op basis van een methodologische analyse van de wetenschappelijke controverses, mogen we concluderen dat de relevantie van onderzoeksvragen *zowel* op het biologische organisatieniveau dat wordt bestudeerd door de ecologie *als* op het biologische organisatieniveau dat wordt bestudeerd door de moleculaire biologie, expliciet moet worden onderbouwd.

De praktische toepassing van de SRQ-benadering had binnen de in Duitsland uitgevoerde TA-studie kunnen leiden tot andere uiteindelijke conclusies en had wellicht kunnen voorkomen dat een deel van de kritische experts de beraadslagingen zou hebben verlaten. De aanspraak van deze TA-studie op het predikaat "participatief technologisch aspectenonderzoek" vindt geen ondersteuning in de werkelijke gang van zaken, omdat een fundamentele en pragmatische methodologische analyse van de betreffende wetenschappelijke controverses ontbrak.

— Hoofdstuk 4 —

In Hoofdstuk 4 wordt het analyse-raamwerk dat werd ontwikkeld in Hoofdstuk 2 en op de proef gesteld in Hoofdstuk 3, gebruikt om een viertal categorieën van in het algemene debat over de bioveiligheidsbeoordeling van gentechnologie steeds terugkerende controversiële kwesties te onderkennen en te evalueren. Er wordt gedemonstreerd dat deze strijdpunten kunnen worden gekenmerkt als kunstmatig gedefinieerd en om die reden een onproductieve ballast voor het debat vormen. De productieve pragmatiek van de SRQ-benadering kan ertoe bijdragen om het debat terug te brengen tot een constructieve waardering van de argumenten die naar voren worden gebracht vanuit het oogpunt van gevaarherkenning. De gepresenteerde analyses dragen ertoe bij om aan te tonen dat de analytische benadering die in deze studie wordt ontwikkeld meer algemeen bruikbaar kan worden gemaakt voor de analyse van individuele beweringen en argumentaties.

De tegenover elkaar staande meningen over de wenselijkheid van bioveiligheidsregelgeving die aan de ene kant is gericht op de *processen* van gentechnologie of aan de andere kant juist is gericht op de *producten* van gentechnologie, worden gereconstrueerd in termen van de relevante onderzoeksvragen die aan de orde worden gesteld door één van beide beleidsopties. Toepassing van de SRQ-benadering maakt duidelijk dat *zowel* ten aanzien van de producten van gentechnologie *als* ten aanzien van de processen van gentechnologie, de relevantie van onderzoeksvragen moet worden beargumenteerd en vergezeld gaat van een wetenschappelijk bewijslast. Dit neemt de onproductieve wederzijdse uitsluiting van de strijdende perspectieven weg en opent de weg naar een constructieve aanvulling vanuit de twee algemene invalshoeken ten behoeve van het oogmerk van gevaarherkenning. Dit wordt aangetoond door middel van een analyse van de methodologische onhelderheid die er bestaat ten aanzien van het in deze discussie gebruikte begrip 'precisie' en ten aanzien van een adequate interpretatie van de biologische rol van transposons.

Vervolgens worden de tegenover elkaar staande standpunten omtrent de wenselijkheid van bioveiligheidsregelgeving die is gebaseerd op een *per-geval-beoordeling* of juist op een *generieke beoordeling*, geanalyseerd door toepassing van de SRQ-benadering. In deze context kan het begrip 'adaptatie' worden geanalyseerd als een mogelijke methodologische valkuil waardoor een controverse kunstmatig in plaats van fundamenteel kan worden. Een hiermee samenhangende moeilijkheid die wordt geanalyseerd is het biologische belang van het ontvangende ecosysteem in de bioveiligheidsbeoordeling. Omdat het alleen zinvol is om deze zorg specifiek te maken in termen van de relevante onderzoeksvragen, kan een gebrek hieraan worden gezien als een recept voor het ontstaan van kunstmatige controverse.

Door de invalshoek te richten op het vinden van de relevante onderzoeksvragen voor het doel van gevaar-herkenning, kan duidelijk worden gemaakt dat *zowel* de overwegingen die liggen vervat in een per-geval-beoordeling *als* de overwegingen die liggen vervat in een generieke-beoordeling kunnen bijdragen aan een verbetering van het inzicht dat nodig is voor het doel van gevaarherkenning.

Een speciale zorg in het debat over de bioveiligheidsbeoordeling van gentechnologie is gelegen in de gevaarherkenning van genetisch veranderde *micro*-organismen. Door middel van toepassing van de SRQ-benadering wordt aangetoond dat een methodologische analyse laat zien dat tenminste een deel van de discussie hieromtrent over kunstmatige kwesties gaat. De aannemelijkheid van beweringen dat genetisch veranderde micro-organismen reeds tot ontwikkeling zijn gekomen in de geschiedenis van de biologische evolutie of dat genetisch veranderde micro-organismen niet schadelijk zullen kunnen zijn omdat er 'geen ruimte' zou zijn in een bestaand ecosysteem, kunnen alleen worden beoordeeld door een specificering van de relevante onderzoeksvragen die hieromtrent gesteld moeten worden vanuit het oogmerk van gevaarherkenning.

Van algemeen belang voor de bioveiligheidsbeoordeling van gentechnologie is een juiste interpretatie van het begrip 'waarschijnlijkheid' binnen een evolutionaire context. Toepassing van de SRQ-benadering laat zien dat het geïmpliceerde begrip van 'onafhankelijke kansen' binnen de biologie specifiek moet worden gemaakt door de relevantie van onderzoeksvragen met het oog op het onderzoeksdoel aan de orde te stellen. Als een voorbeeld van een 'kunstmatige controverse' over de waarschijnlijkheid van het optreden van ongewenste effecten binnen de context van de bioveiligheidsbeoordeling van gentechnologie, wordt de bewering dat er een mechanisme van 'voorafgaande waarschuwing' zou bestaan binnen de biologische werkelijkheid geanalyseerd en gereconstrueerd in termen van de onderzoeksvragen die als relevant moeten worden beschouwd om deze bewering te ondersteunen.

— Hoofdstuk 5 —

Binnen de context van het bioveiligheidsbeleid is het begrip 'vertrouwdheid' ('familiarity') geïntroduceerd als poging om een pragmatische basis te vinden voor een wetenschappelijke onderbouwing van de bioveiligheidsbeoordeling. Deze poging wordt ondermijnd door het feit dat een bruikbare wetenschappelijke definitie van 'vertrouwdheid' tot op heden heeft ontbroken binnen deze context. In Hoofdstuk 5 wordt aangetoond dat de SRQ-benadering kan worden toegepast als een praktische methodologische basis voor een wetenschappelijke definitie van het begrip 'vertrouwdheid'.

Vertrouwdheid kan alleen specifiek en bruikbaar worden gemaakt binnen de context van wetenschappelijke deskundigheid voor beleidsdoeleinden door het expliciet te maken in termen van de verzameling van onderzoeksvragen die als relevant wordt beschouwd om aanspraak te kunnen maken op voldoende bekendheid met het onderwerp van studie. Ongespecificeerd gebruik van dit begrip vormt een uitnodiging voor het ontstaan van een kunstmatig inzicht in de wetenschappelijke voorwaarden voor een betrouwbare gevaarherkenning en bioveiligheidsbeoordeling. De voorgestelde pragmatische benadering stelt ons in staat om een internationaal onderzoeksproject te ontwerpen en te organiseren met het oogmerk om de wetenschappelijke vereisten voor een aanspraak op 'vertrouwdheid' te specificeren in termen van de onderzoeksvragen die als relevant beschouwd moeten worden voor een afdoende bestudering van het onderwerp.

Het begrip 'horizontale gen-overdracht' en het gebruik ervan in discussies over de bioveiligheidsbeoordeling van gentechnologie wordt geanalyseerd om te demonstreren dat het begrip 'vertrouwdheid' alleen zinvol gebruikt kan worden binnen de context van bioveiligheidsbeoordeling en bioveiligheidsbeleid wanneer het operationeel kan worden gemaakt in termen van specifieke relevante onderzoeksvragen.

De SRQ-benadering maakt een *dynamische* en *modulaire* procedure mogelijk in de ontwikkeling van een omschrijving van 'vertrouwdheid'. Controverses die overal op de wereld bestaan over de bioveiligheid van gentechnologie kunnen aldus productief worden gemaakt door ze te interpreteren als evenzovele bronnen van onderzoeksvragen die mogelijk relevant zijn voor het gekozen onderzoeksdoel. Door de aandacht te richten op de onderscheiden vraagstellingen kunnen verschillen van mening vruchtbaar worden gemaakt in plaats van voortslepend en teleurstellend.

Een praktisch voordeel van deze benadering ligt in het feit dat het mogelijk wordt gemaakt om te werken aan de ontwikkeling, als een internationaal proces van vereende krachten, van een *dynamische* en *modulaire* catologus van onderzoeksvragen die tezamen een basis kunnen geven aan voldoende 'vertrouwdheid'. Een dergelijke catalogus kan een vraagbaak worden voor de specificering van zulke aanmeldings- en monitoring-vereisten als nu bijvoorbeeld worden uitgewerkt door de regelgeving in *Annex II* van *EU Council Directive* 90/220. Aldus kan deze benadering ook ondersteuning bieden aan nationale adviesraden van de Competente Autoriteiten op het gebied van de bioveiligheidsbeoordeling, zoals de Nederlandse COGEM (*Commissie voor Genetische Modificatie*), voor welke het op dit moment in sommige gevallen niet duidelijk is wat een voldoende wetenschappelijke basis zou kunnen zijn om te besluiten tegen verlenen van een vergunning voor veldproeven of markttoelating van een genetisch veranderd organisme.

Als een praktisch raamwerk voor de ontwikkeling van een dergelijke catalogus, is een voorlopige definitie nodig van wat kan worden verstaan onder het begrip 'gevaar' om aan te geven welke aspecten van 'vertrouwdheid' bestudeerd moeten worden voor het onderzoeksdoel van gevaarherkenning.

De voorgestelde definitie luidt: "Een '*genetisch veranderd organisme* vertegenwoordigt een *gevaar*' wanneer 'het een *Agent P* draagt, die een *Effect Q* kan hebben dat als onwenselijk wordt beschouwd in *Context R*, op een *Getroffene S*, via een *Mechanisme T*, in een *Omgeving X*, als gevolg van een *Toepassing Z*'." Hiermee wordt een zevental predikaten of aandachtspunten aan de orde gesteld die mogelijk relevant zijn voor het ontwerp van adequate onderzoekprogramma's en monitoringstrategieën voor bioveiligheidsbeoordeling van gentechnologie.

— Hoofdstuk 6 —

In Hoofdstuk 6 worden de consequenties van de bevindingen van dit onderzoek voor de praktische status van wetenschappelijke deskundigheid bij elkaar gebracht en geëvalueerd. De SRQ-benadering impliceert een fundamentele verantwoordelijkheid voor wetenschappelijke experts.

Wanneer politici op basis van hun democratische legitimering besluiten om bepaalde mogelijk relevante vraagstellingen niet aan de orde te stellen, dan is dit een politieke beslissing in de daartoe geëigende context. Wanneer wetenschappers een dergelijk besluit nemen, dan is dat een politieke beslissing in een *oneigenlijke* context.

De demarcatie van *wetenschappelijke* en *transwetenschappelijke* vragen binnen de bioveiligheidsbeoordeling is een verantwoordelijkheid van wetenschappers omdat zij in de beste positie verkeren om te beoordelen welke vragen wel en welke vragen niet door de wetenschap beantwoord (zouden) kunnen worden.

Politieke en ethische afwegingen kunnen niet worden gemaakt zonder een voldoende inzicht in de methodologische status van de wetenschappelijke beweringen die deel uitmaken van de afweging. Een wetenschappelijke expert draagt, *als wetenschapper*, een verantwoordelijkheid om de kwaliteit en de relevantie van empirische gegevens te waarborgen. Wetenschap die op voorzorg is gericht wijdt een speciale opmerkzaamheid aan de beperkingen van wetenschappelijke beweringen en poogt een bestaand gebrek aan informatie aan het licht te brengen met de hulp van kritische wetenschappelijke methodologie. Zonder deze methodologische toets, die de essentie vormt van wetenschappelijk onderzoek, is er nauwelijks een manier voorstelbaar om het internationaal geaccepteerde *voorzorgsprincipe* daadwerkelijk toepasbaar te maken. Methodologische valkuilen zoals die bijvoorbeeld liggen in de interpretatie van statistische resultaten die zijn geproduceerd met het oog op gevaarherkenning kunnen eigenlijk alleen herkend en vermeden worden door de wetenschappelijke experts die betrokken zijn bij de uitdaging van bioveiligheidsbeoordeling. Een belangrijke taak van wetenschap is om ons te behoeden voor de gedachte dat we meer weten dan in werkelijkheid het geval is. In de epiloog van deze studie worden de belangrijkste conclusies van het onderzoek bijeengebracht en geplaatst tegen de achtergrond van de *leerprocessen* die een noodzakelijke voorwaarde zijn voor een zorgvuldige bioveiligheidsbeoordeling van gentechnologie.

Index